Processing and Properties for Powder Metallurgy Composites

Processing and Properties for Powder Metallurgy Composites

Proceedings of a symposium sponsored by the P/M Committee of TMS/AIME held at the Annual Meeting of the Metallurgical Society in Denver, Colorado, February 1987.

Edited by

P. Kumar
Cabot Corporation
Boyertown, PA

K. Vedula
Case Western Reserve University
Cleveland, OH

A. Ritter
General Electric Corporation
Schenectady, NY

A Publication of The Metallurgical Society, Inc.
420 Commonwealth Drive
Warrendale, Pennsylvania 15086
(412) 776-9000

Printed in the United States of America.
Library of Congress Catalogue Number 87-43115
ISBN NUMBER 0-87339-034-2

Preface

Powder Metallurgy Processing of Composites was considered an ideal topic for the symposium at the Annual Meeting of The Metallurgical Society in February 1987 because of the commercial acceptability of metal matrix composites, and the advent of the intermetallic matrix composites as potential materials of the future. This monograph includes papers presented at this symposium. Papers addressing both the fabrication via powder metallurgy techniques and the properties of composites are included.

This book is organized by topic, with the first section on the processing of powder metallurgy composites, and the second section emphasizing composite properties.

Processing methods usually consist of one of the following: pressing and sintering, wrought-processing and coatings. All three types of methods offer the required flexibility to obtain a range of microstructure and, therefore, to have a better control in materials design. This important concept of flexibility in materials design is emphasized in papers by Lawley and Vedula. Papers by Kaysser and his co-workers describe the production of heavy metal alloys by liquid phase sintering techniques.

A glimpse of the impact of emerging technologies, such as low pressure plasma deposition (LPPD), liquid dynamic composition and spray-forming on the fabrication of composites, is provided by Jackson and his co-workers.

This is followed by papers which are concerned with properties of composites. Work on the microstructure and properties of plasma-sprayed steel/carbide composites by Ritter, Jackson and Wright is presented; followed by Dixon and Wright's research on creep deformation in cemented carbides and the evaluation of these materials by stress relaxation test methods.

Trybus and her co-authors discuss the fabrication of a new material (Cu-Nb composite) via wrought processing. P/M is preferred over ingot metallurgy due to differences in melting points of Cu and Nb. However, high oxygen in niobium powder might cause degradation.

Several papers on aluminum matrix composites reflect the current popularity of these alloys. Harris and Wawner investigated the interrelation of microstructure and tensile strength in Al-SiC composites processed under different conditions. Lewandowski and his co-workers have assessed the effects of microstructure on the fracture mechanisms in this type of composite. The unique properties of composite microstructures of modified aluminum alloys with a variety of dispersoids are addressed by Murty and Koczak. The experimental composites, such as Al/B_4C and Al/Al_2O_3, are discussed by Rack.

The complexity of microstructural and material response to the unique processing methods provided by powder metallurgy is apparent in these papers.

The editors would like to take this opportunity to thank the authors for their contributions and the TMS staff members for this assistance.

P. Kumar
K. Vedula
A. Ritter

Table of Contents

COMPOSITE STRUCTURES USING FLEXIBILITY OF POWDER METALLURGY

K. Vedula

Associate Professor
Department of Materials Science and Engineering
Case Western Reserve University
Cleveland, Ohio 44106

Abstract

A unique advantage of powder metallurgy is the ability to tailor microstructures in the form of composites of different materials. Although this concept has been exploited in several instances in the past (e.g. in cemented carbides), a variety of new concepts have evolved in recent years. This paper will briefly review some of these concepts, with emphasis on macro-composites with the help of specific examples.

Processing and Properties for Powder Metallurgy Composites
Edited by P. Kumar, K. Vedula and A. Ritter
The Metallurgical Society, 1988

Introduction

Powder metallurgy provides a flexible process for manufacturing composites with tailored microstructures. With the wide range of powder variables and process variables available, it is possible to design microstructures in such a manner as to obtain specific properties at specific locations in the component. This concept can be applied at the microscopic level (referred to in this paper as micro-composites) as well as at the macroscopic level (referred to in this paper as macro-composites).

Micro-Composites

The microscopic tailoring of microstructures using immiscible powders is already an ancient technology and has been very successfully applied to sintered cemented carbides in which hard carbide phases are embedded in a softer matrix and used for a variety of cutting and drilling applications. Similar examples can be found throughout a range of applications including electrical contacts, bearing applications, high temperature structural alloys etc. and a review of these is not within the scope of this paper.

The recent renewed interest in composites for structural applications has opened up a wide range of new processing opportunities for powder metallurgy. A typical processing sequence for a reinforced composite is illustrated in Figure 1. Fiber reinforcement is usually restricted to short fibers due to the limitations of the process. Powder processing is much more appropriate for particulate composites. A variety of combinations of metals, ceramics and polymers are already being processed using these techniques. The majority of the papers in this symposium will address micro-composites and, hence, in this paper the unique opportunities of macro-composites will be emphasized.

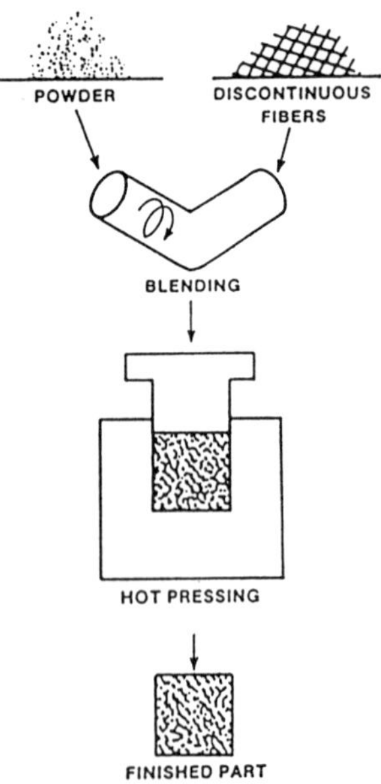

Figure 1 Schematic of a typical processing sequence for micro-composites.

Macro-Composites

A) Diffusion Bonded

Several programs over the past several years, sponsored by the Air Force, by NASA Lewis Research Center as well as by private industry have demonstrated the feasibility of diffusion bonding different parts of complex components using a P/M diffusion bonding approach. P/M dual property turbine wheels for small gas turbine engines is one example. These wheels include powder metal hubs in conjunction with cast blading (1). High temperature strength, which is a requirement for the blade, is normally achieved via a coarse-grained structure or by special processing such as directional solidification. Casting techniques are ideally suited for this purpose. On the other hand, the important considerations for the disk material are moderate temperature strength coupled with fatigue resistance. Such a combination is achieved in a fine-grained structure which is readily attained by a powder metal product. A hybrid of the two microstructures, hence, provide the best possible component.

Figure 2 illustrates the Avco concept for such a dual property wheel. The metallurgical bonding of a directionally solidified blade to a P/M hub is achieved by HIP consolidation using a ceramic mold approach. The test results for the HIP bond exceeded the design predictions in both the hub section as well as the blade-hub interface location.

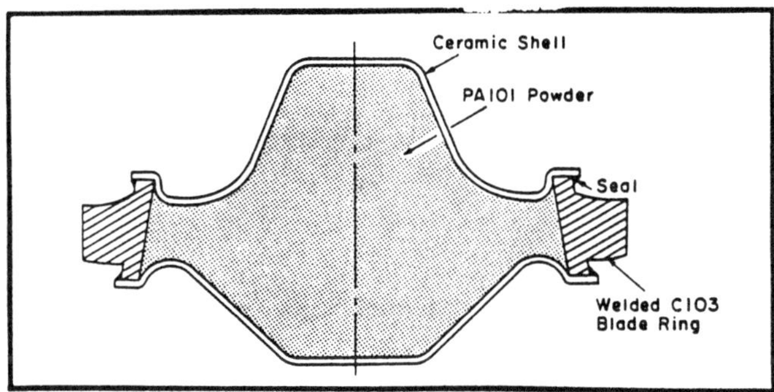

Figure 2 The AVCO Concept of the P/M dual property turbine wheel (1).

HIP diffusion bonding of P/M alloys for composite land-based gas turbine buckets is another example of a successful macro-composite (2). Using this technique, airfoil vanes can be made of directionally solidified (DS) nickel base superalloys for optimizing creep/rupture properties while dovetail regions would use P/M superalloys for high tensile strength and ductility. The limiting item for this technology is the critical airfoil to dovetail bondline. Figure 3 illustrates the process involved. Careful bonding techniques and proper heat treating have been shown to result in metallurgical bonds having strengths and ductilities greater than the DS materials. This is dependant on minimizing the formation of brittle second phases along the interface. Maximum bond strength is obtained when the fracture is forced off the bondline and into the weaker cast material.

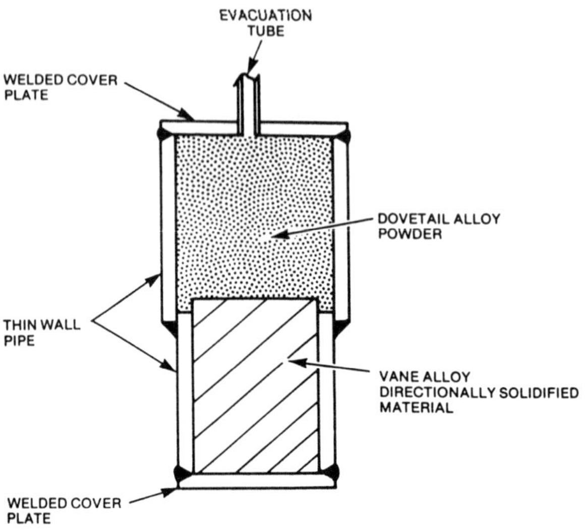

Figure 3 HIP can design for diffusion bonding DS alloys to consolidated powder (2).

B) Sinterbonded

Sinterbonding can be used to produce P/M composite parts which are made up of metal powder and solid metal. The advantages of this technique are that complex shapes can be made more easily, high mechanical strength can be obtained at required locations and parts which contain some porous regions can be easily fabricated, by selecting appropriate combinations of powder and solid metal.

An example of such a composite system is illustrated in Figure 4 (3). The powder and the solid steel are placed concentrically in a closed die and formed together. The two portions are subsequently bonded during the sintering operation. Of particular interest is that the densification and resulting volume change associated with sintering can be used to improve the bonding between the P/M and the solid portions. The bond strengths of such composite compacts can also be improved by repressing and resintering. Suitable tailoring of the materials can vastly improve the bond strength. For instance, in this example of the powder inside the ring (Figure 4b), an outward bonding stress can be generated by using admixed powders of iron and copper which increase in dimensions during sintering.

Density distributions in the P/M portion of such composites can be fairly uniform in spite of the large height to thickness ratios. Furthermore, increased surface roughness of the solid metal part can improve the bonding between the powder and the metal portion by providing for more surface area of contact.

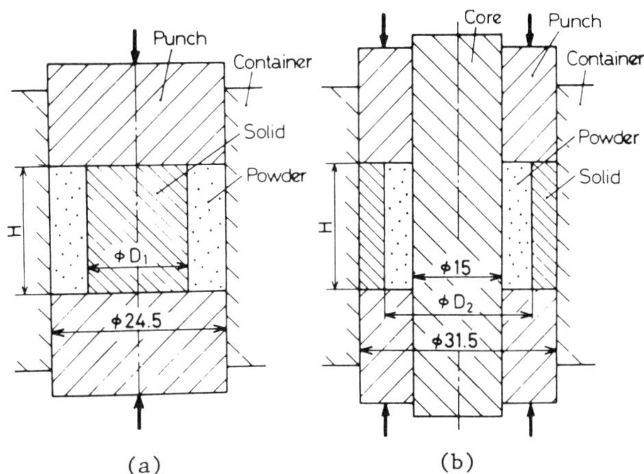

Figure 4 Schematic representation of set-up for two types of composites
a) powder outside a solid metal cylindrical part and
b) powder inside a solid metal ring (3).

P/M diffusion bonding during the sintering process is another useful technique which has been developed to bind multiple green compacts into a single component. This is more economical than other techniques for obtaining complex configurations of components. The different components are once again tailored for their mechanical shrinkage characteristics. Several examples of gear assemblies produced by this technique using inner and outer parts are given by Sakai and Asaka (4). To obtain high bond strengths, the inner material must have a larger thermal expansion than the outer material during sintering. Additions of alloying elements such as Ni, Cu, C and P to an iron matrix can be used to control the thermal expansion and shrinkage characteristics, thereby increasing the pressure for bonding at the interface.

C) Forming of Sintered Composites

Macrocomposites can also be prepared by secondary processing of sintered composites in order to improve densities and properties. An example of this is provided by an electrical component by Grosse (5) in which a combination of high resistance to welding at the switching side of the electrical contacts along with high welding strength at the contact carrier is required. Components must, therefore, be made from different materials in the form of sintered composites which can be further densified by a forming step.

Another example of a formed sintered composite is the X-ray tube target developed by General Electric Company which involves a two-layered compact which is sintered and subsequently forged to obtain full density. The macro-composite illustrated in Figure 5 has an interface between the W and Mo portion of the target. A recent study by CWRU and General Electric Co. (6) has shown that differences in the compaction characteristics of the two powders frequently cause delaminations at the interface in the green state as well as during subsequent sintering and forging operations. The critical aspect of the green strength at the interface was addressed, and it was demonstrated that control of the processing variable such as the powder size, powder composition and compaction pressure can result in a good green bond between the two portions which can then withstand the sintering and forming operations successfully.

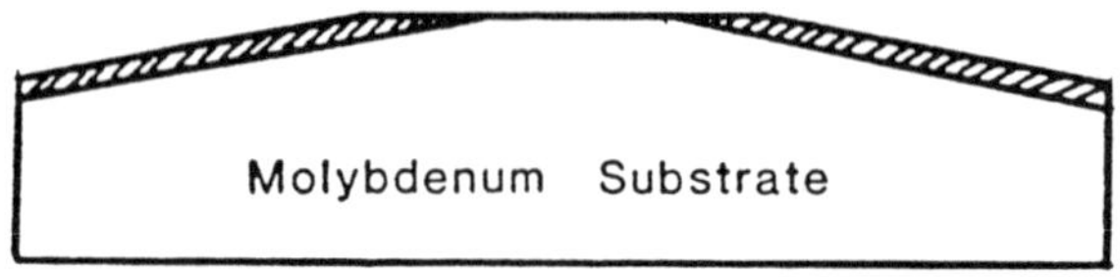

Figure 5 Schematic of the W-Mo composite compact G.E. X-ray tube target (6).

Conclusions

In summary, it must be emphasized that powder metallurgical processing offers a wide range of control and flexibility which, if properly understood and utilized, can contribute to tremendous gains in the processing of complex parts.

References

1. J. H. Moll, J. H. Schwertz and V. K. Chandok, Progress in Powder Metallurgy, 1981, ed. J. M. Capus, D. L. Dyke, 1981, p. 303.

2. L. G. Peterson, D. Hrencecin, A. Ritter and Nathan Lewis, Progress in Powder Metallurgy, 1985, p. 561.

3. T. Tabata and S. Masaki, Modern Developments in Powder Metallurgy, Vol. 16, 1984. p. 181.

4. J. Sakai and K. Asaka, Horizons of Powder Metallurgy, Proceedings of the 1986 International P/M Conference, 1986, p. 717.

5. J. Grosse, ibid, p. 317.

6. S. Strothers and K. Vedula, International Journal of Powder Metallurgy. Vol. 22, No. 4, 1986, p. 227.

HIERARCHICAL MATERIALS STRUCTURES

A. Lawley and M.J. Koczak

Department of Materials Engineering
Drexel University
Philadelphia, PA 19104

Abstract

It is now possible to deşign metallic alloy systems possessing hierarchical structures utilizing singly or in combination, the several technologies of powder metallurgy processing. For example atomization, with its inherent component of rapid solidification, applied to aluminum allloys containing transition metal elements, provides a prealloyed powder with a fine-scale intermetallic dispersoid ~0.2 μm dia. The mechanical alloying of this powder further introduces oxides and carbides at a scale of about 30 nm. After consolidation via hot extrusion, such dual scale structures exhibit remarkable levels of strength, strength retention above ambient, and microstructural stability. Recent observations and results for Al-Fe-Ni and Al-Ti are examined, in particular hardness, strength, and creep response. These hierarchical structures appear suitable for structural applications at temperatures well above the limits for conventional dispersion strengthened aluminum alloys.

Processing and Properties for Powder Metallurgy Composites
Edited by P. Kumar, K. Vedula and A. Ritter
The Metallurgical Society, 1988

Introduction

Today's complex and engineered materials systems are invariably composite or multiphase in nature. They often comprise different structural scales, ranging from the macroscopic to the nanoscopic, may perform several different functions, and may even encompass varying levels of architectural organization. If the latter condition prevails, the materials may accurately be described as hierarchical materials structures (HMS).

Hierarchical systems abound in nature. For example macromolecular building blocks in plants and animals involve several levels of highly organized hierarchical architectures. Such multilevel architectures provide the key to the unique directional structure-property relations in these natural systems and their resistance to stress-induced damage. Similarly, the hierarchical construction of collagen fibrous tissues in animals and humans extends across the entire range of the structural scale. Thus, a basic understanding of these different levels of microstructural organization is required in order to account for and explain the structural-mechanical functions of these natural biocomposites.

Multilevel hierarchical architectures are now becoming more common in man-made (engineered) polymers and inorganic multicomponent materials. Examples include flow-oriented thermotropic liquid crystals, polymer composites, blends and alloys, CVD/PVD multilayer structures, ion implanted and molecular beam epitaxial materials, rapidly solidified alloys, and mechanically alloyed materials. The multifunctional hierarchical systems utilizing these materials include integrated circuits, optoelectronic circuits, quantum-well superlattices and heterofunctions, highly integrated solid state sensors, and active polymer membranes.

In this contribution we focus on the hierarchical aspects of the design and synthesis (processing) of two complex aluminum alloys. It is shown that by a combination of powder metallurgy (P/M) processing and mechanical alloying (MA), unique microstructures evolve, and in turn these alloys exhibit unusually attractive mechanical properties, particularly above ambient temperature.

Metallic HMS via Particulate Processing

In metallic systems, the scale of size refers to microstructural constituents. Primary examples are the grain size of the matrix and the size of any precipitates or dispersoids.

Particulate processing offers several approaches to the synthesis of HMS. The intrinsic component of rapid solidification in the several modes of atomization gives rise to HMS in the powder product. Subsequent mechanical alloying further refines the size range of microstructural features present in the powder product. Other particulate processing methods that result in HMS are Osprey™ spray deposition and low pressure plasma deposition.

Aluminum Alloy HMS

There are several inherent limitations to achieving property and performance goals in aluminum alloys via ingot metallurgy (I/M) and this has stimulated research into powder metallurgy as a processing alternative (1,2). Of particular interest is the technology of rapid solidification processing of aluminum alloys, since it has been demonstrated that this approach provides enhanced alloying flexibility, and results in refined fine-scale homogeneous microstructures with minimal attendant solute segregation, compared with I/M alloys (3-6).

Several binary and ternary aluminum alloys have been investigated (3,5,6,7-17). The selection of the second and third alloying elements has been based primarily on the following criteria: high liquid solubility in aluminum, a low solid solubility in aluminum, and a low rate of solid state diffusion in aluminum. The first criterion permits large alloying additions to be made to the melt, while the second criterion ensures almost complete precipitation on cooling to form a high volume fraction of second phase particles (e.g. intermetallics). If the particles are strong and non- deformable, dislocations will be forced to bow between them during deformation. For a given particle size, an increase in the volume fraction of the second phase particles decreases inter- particle spacing, which in turn increases strength. Further, if the intermetallic particles have high elastic moduli, then a material containing a high volume fraction of such particles will have a high Young's modulus. The upper temperature limit is set by the stability or resistance to coarsening of the dispersed particles. Hence, it is necessary that the alloying elements selected have low rates of diffusion in solid aluminum. This third criterion minimizes the rate of coarsening of the dispersed particles at elevated temperatures.

On the basis of these criteria, the alloying elements selected in the development of a new class of high temperature aluminum alloys are the transition elements Cr, Mn, Fe, Ni and Co, and the rare earth element Ce. Alloy systems that show distinct promise in achieving property and performance goals at elevated temperature are Al-Fe-Ni, Al-Fe-Ce and Al-Fe-Mo. The fine-scale intermetallic dispersoid, uniformly distributed throughout the aluminum matrix, is achieved by atomization of the molten alloy with its intrinsic component of rapid solidification. Subsequently the powder is hot consolidated to full density.

A fine scale uniform distribution of dispersoids can also be introduced into an alloy by mechanical alloying (MA) (18-23). MA results in a fine-scale dispersion of Al_2O_3 since the process breaks up oxide films present on powder particle surfaces, and it also introduces small dispersoids of Al_4C. The source of carbon is the process-control agent added to prevent excessive welding during MA.

(a) Al-Fe-Ni

Processing details for the P/M and MA Al-Fe-Ni alloys are given in Appendix I. A representative microstructure of as-extruded MA Al-Fe-Ni is shown in Figure 1(a). Fine oxides and carbides are distributed uniformly within the aluminum matrix between the $FeNiAl_9$ dispersoids. The average intermetallic particle size is about 0.18 μm whereas the size of the oxide/carbide dispersion is about 30 nm.

Microstructural stability of the HMS is illustrated in Figure 1(b). After 288 hours at 450°C only a limited coarsening of the microstructure is observed, cf. Figures 1(a) and 1(b). Even more compelling is the microstructural stability exhibited after 50 hours at 550°C and 610°C, Figures 2(a) and 2(b) respectively. The influence of the nanometer scale dispersoid of oxides/carbides in promoting elevated temperature microstructural stability is demonstrated in Figure 3. In comparison with the MA material (Figures 3(a) and 3(b)), the non MA material (Figures 3(c) and 3(d)) reveals coarsening of the $FeNiAl_9$ intermetallic and the subgrain size is larger.

Consistent with the enhanced elevated temperature microstructural stability, the corresponding mechanical properties are improved. This is illustrated in terms of hardness in Figures 4 and 5 for isochronal and isothermal conditions, respectively. Codes (a), (b) and (c) refer to the prior processing conditions and are explained in Appendix I. From Figure 4 it is seen that after exposure above 400°C, the ambient temperature hardness of the non-MA material decreases precipitously. In comparison, the MA material exhibits a temperature advantage of about 100°C. Even above 500°C, the ambient temperature hardness of the MA material decreases gradually with increasing temperature of exposure. Under isothermal conditions, the drop in ambient temperature hardness after 624 hours at 450°C is only about 10% in the MA materials, Figure 5. Without MA, the corresponding decrease is about 70%.

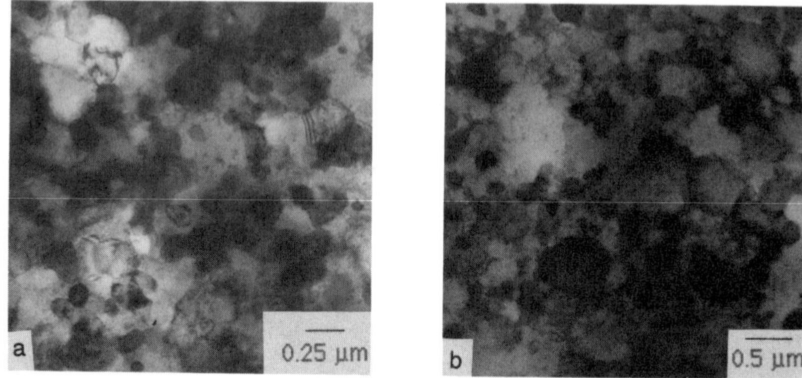

Figure 1: Microstructure (TEM) of MA Al-Fe-Ni alloy; a-processing. (a) As-extruded, (b) 450°C/288 hrs.

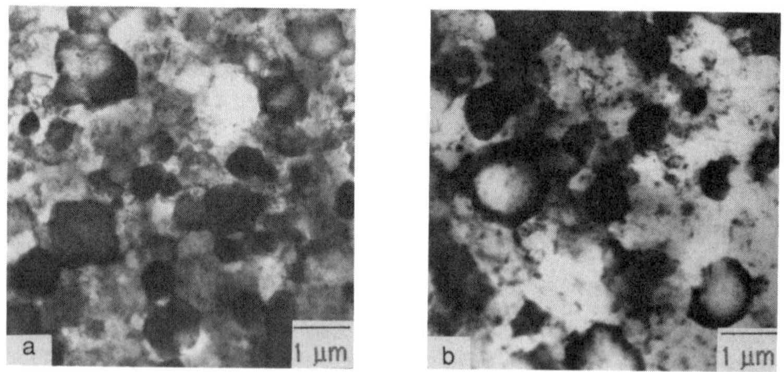

Figure 2: Microstructure (TEM) of MA Al-Fe-Ni alloy; b-processing. (a) 550°C/50 hrs., (b) 610°C/50 hrs.

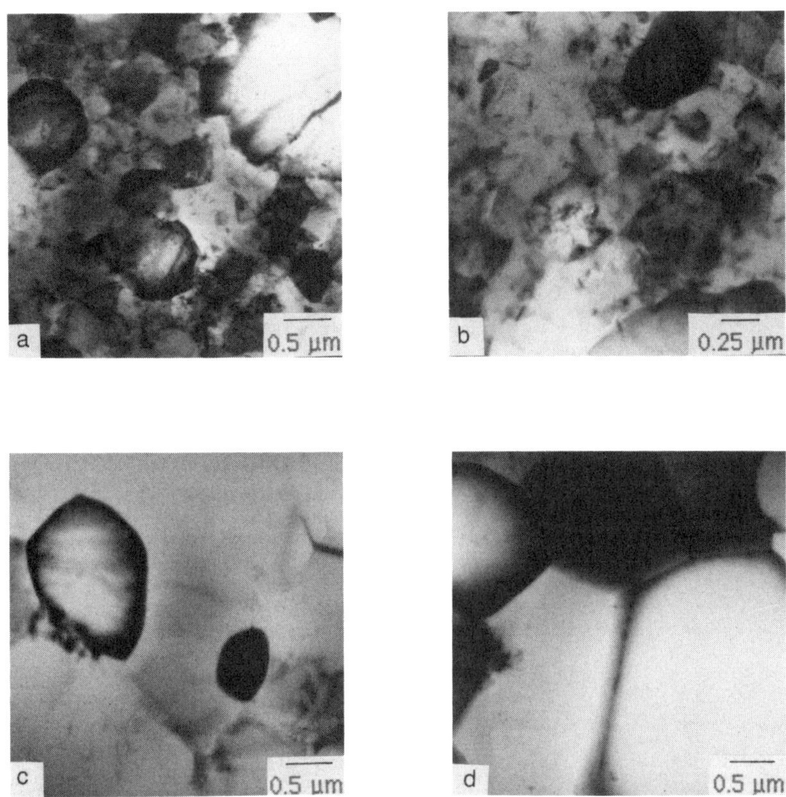

Figure 3: Microstructure (TEM) of Al-Fe-Ni alloy; b-processing. (a) MA/550°C/50 hrs., (b) MA/550°C/288 hrs., (c) Non-MA/550°C/50 hrs., (d) Non-MA/550°C/288 hrs.

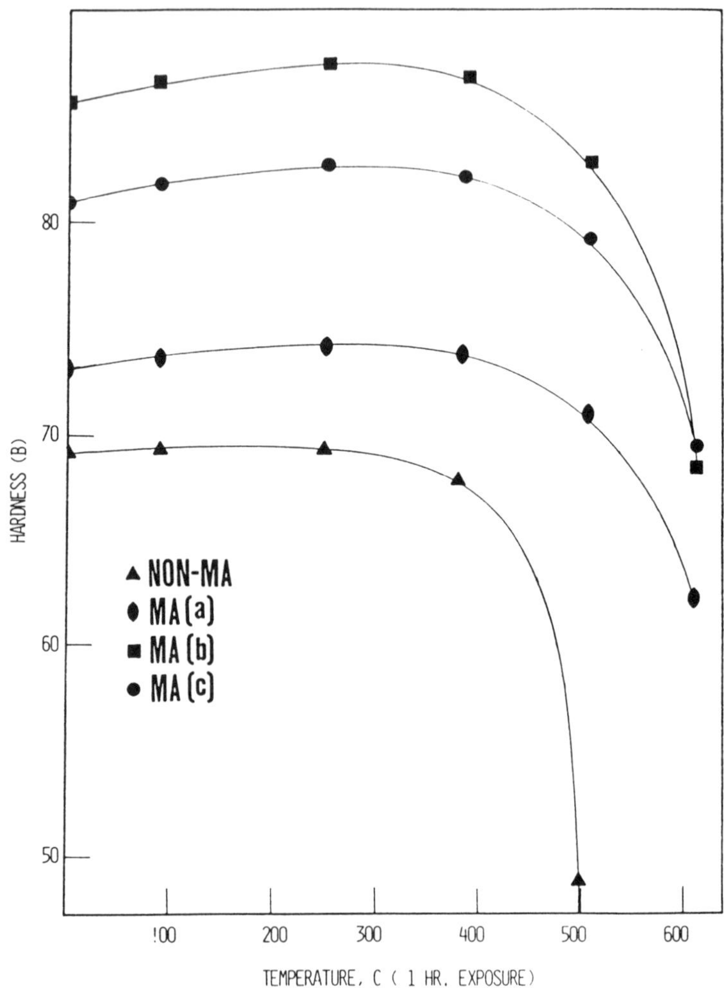

Figure 4: Room temperature hardness of Al-Fe-Ni alloy after isochronal elevated temperature exposure.

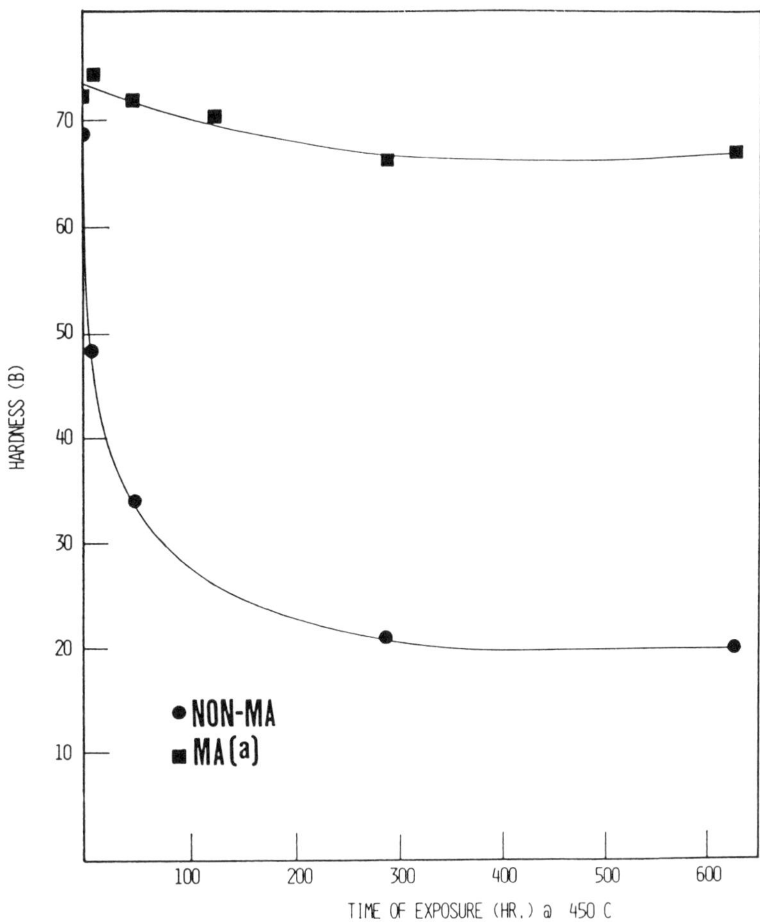

Figure 5: Comparison of room temperature hardness of non-MA Al-Fe-Ni and MA Al-Fe-Ni; isothermal exposure at 450°C.

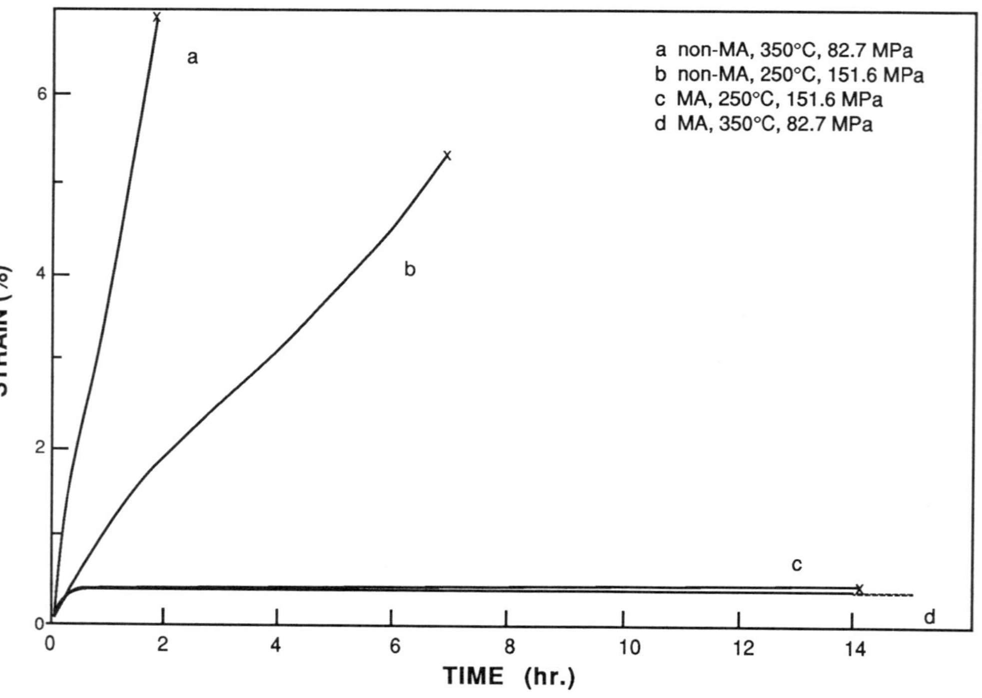

Figure 6: Comparison of creep response of non-MA and MA Al-Fe-Ni alloy; a-processing.

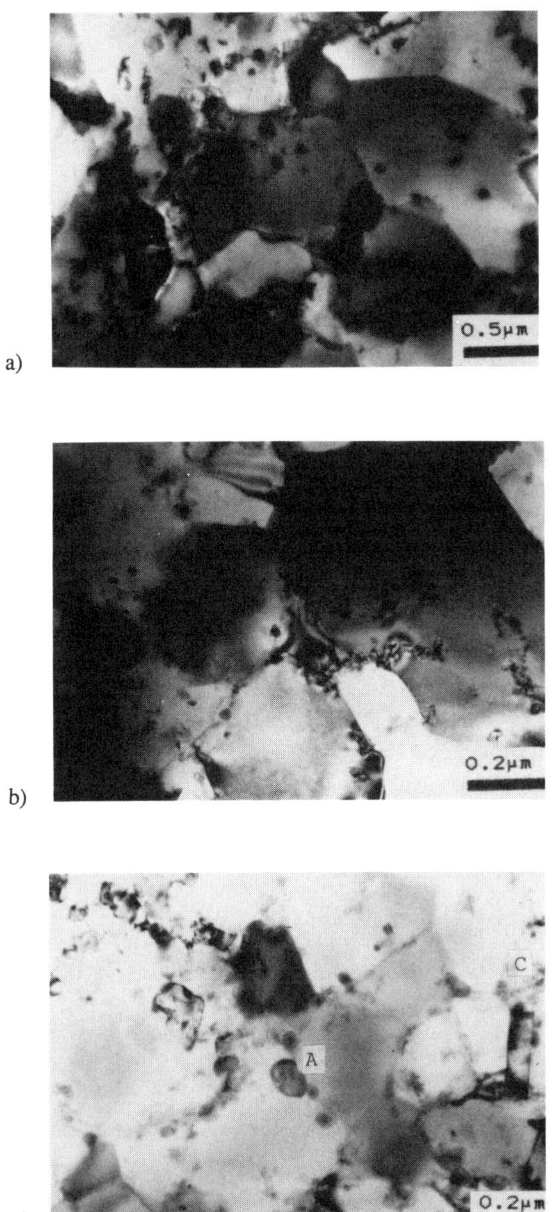

Figure 7: Microstructure (TEM) of Al-Ti alloy. (a) Non-MA/AT6, (b) MA/AM6, (c) MA/MA6 Al_3Ti and Al_4C_3/Al_2O_3 marked A and C respectively.

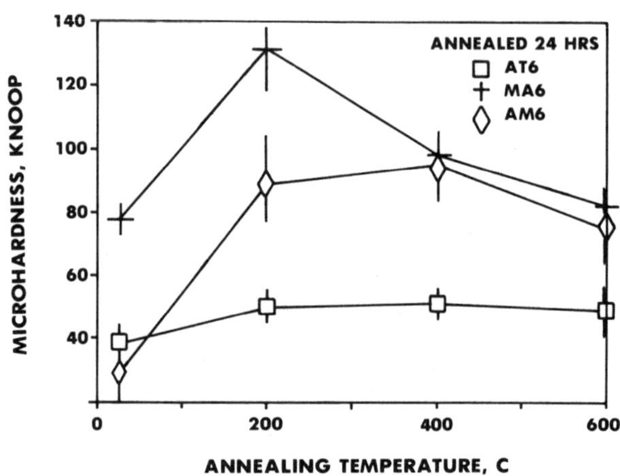

Figure 8: Room temperature microhardness of Al-Ti non-MA and MA alloys after isochronal elevated temperature exposure.

Similarly, creep resistance is enhanced significantly by the presence of the fine scale oxides/carbides. A comparison of some creep curves of the non-MA and MA material under identical conditions of temperature and stress is made in Figure 6.

The increase in strength brought about by MA is accompanied by a decrease in ductility, as measured by strain to failure. Quantitatively, MA increases ambient temperature yield strength by about 77%, but there is a decrease in ductility of about 85%.

(b) Al-Ti

Processing details for the P/M and MA Al-Ti alloys are given in Appendix II. Representative micrographs of as-extruded Al-Ti are shown in Figure 7. In the absence of MA, the microstructure consists of Al_3Ti particles dispersed in the aluminum matrix, Figure 7(a). After MA, a fine dispersion (~0.01μm) of Al_2O_3 and Al_4C_3 is present, located preferentially at the grain boundaries, Figures 7(b) and 6(c). The grain size (~0.4 μm) and distribution of the Al_3Ti intermetallic phase is similar with or without MA.

The influence of the submicron-scale dispersion of Al_2O_3 and Al_4C_3 on microstructural stability above ambient is illustrated in Figure 8 for isochronal annealing. Hardness data are measured at ambient. It is seen that the microhardness of the non-MA material (code AT6) is significantly lower than that of the two MA materials (codes MA6 and AM6).

There is significant increments in yield and tensile strength caused by the fine-scale dispersion of oxides and carbides. For the Al-6 w/o Ti alloy, ambient temperature yield strength increases from 180 MPa to about 322 MPa as a result of MA. There is, however, an accompanying decrease in ductility (elongation) from 21% in the non-MA material to about 8-9% after MA.

Conclusions

1. Singly or in combination, powder metallurgy processing technologies can be utilized to fabricate HMS.

2. In aluminum-base alloys the scale of microstructural features ranges from the nanometer to the micron level.

3. The presence of nanometer scale oxides and carbides enhances strength and strength retention at elevated temperatures, but at the expense of ductility.

Acknowledgments

The authors acknowledge program support from the Air Force Office of Scientific Research, Washington, DC and the Naval Air Development Center, Warminster, PA. Sallah Ezz and William E. Frazier assisted with the studies on Al-Fe-Ni and Al-Ti, respectively.

References

1. S.J. Savage and F.H. Froes, J. of Metals, Vol. 36, 4, p. 20, 1984.

2. A. Lawley, J. of Metals, Vol. 38, 8, p. 15, 1986.

3. W.M. Griffith, R.E. Sanders, Jr., and G.J. Hildeman, in High Strength Powder Metallurgy Aluminum Alloys, Edited by:: M.J. Koczak and G.J. Hildeman, the Metallurgical Society of AIME, Warrendale, PA, p. 209, 1982.

4. N.J. Grant, ibid., reference number 3, p. 3.

5. Rapidly Solidified Powder Aluminum Alloys, Edited by: M.E. Fine and E.A. Starke, Jr., ASTM Special Technical Publication 890, Am. Soc. for Testing and Materials, Philadelphia, PA, 1986.

6. High Strength Powder Metallurgy Aluminum Alloys II, Edited by: G.J. Hildeman and M.J. Koczak, The Metallurgical Society, Inc., Warrendale, PA, 1986.

7. D.L. Yaney, J.C. Gibeling and W.D. Nix, in Strength of Metals and Alloys, Edited by: H.J. McQueen, J.P. Bailon, J.I. Dickson, J.J. Jonas and M.G. Akben, Pergamon Press, Oxford, p. 887, 1986.

8. Y. Kim and W.M. Griffith, ibid., reference number 5, p. 485.

9. D.L. Yaney and W.D. Nix, Met. Trans., Vol. 18A, p. 893, 1987.

10. L. Angers, M.E. Fine and J.R. Weertman, Met. Trans., Vol. 18A, p. 555, 1987.

11. Aluminum Alloys: Their Physical and Mechanical Properties, Edited by: E.A. Starke, Jr., and T.H. Sanders, Jr., The Chameleon Press, London, 1986.

12. Y. Kim, W.M. Griffith and F.H. Froes, J. of Metals, Vol. 32, 8, p. 27, 1985.

13. M.K. Premkumar, A. Lawley and M.J. Koczak, in Modern Developments in Powder Metallurgy, Edited by:: E.N. Aqua and C.I. Whitman, Metal Powder Industries Federation, Princeton, NJ, U.S.A., Vol. 16, p. 467, 1985.

14. M.K. Premkumar, "A Fundamental Study of P/M Processed Elevated Temperature Aluminum Alloys", Ph.D. Dissertation, Department of Materials Engineering, Drexel University, Philadelphia, PA, 1985.

15. M.K. Premkumar, M.J. Koczak and A. Lawley, ibid., reference number 6, p. 265.

16. S.S. Ezz, M.J. Koczak, A. Lawley and M.K. Premkumar, ibid., reference number 6, p. 287.

17. S.S. Ezz, A. Lawley and M.J. Koczak, ibid., reference number 11, p. 1013.

18. J.S. Benjamin, Met. Trans., Vol. 1A, p. 2943, 1970.

19. J.S. Benjamin, Scientific American, Vol. 234, 5, p. 40, 1976.

20. J.S. Benjamin and M.J. Bomford, Met. Trans., Vol. 8A, p. 1301, 1977.

21. G. Jangg, F. Kutner and G. Korb, Powder Metall. Int., Vol. 9, p. 24, 1977.

22. P.S. Gilman and W.D. Nix, Met. Trans., Vol. 12A, p. 813, 1981.

23. R.F. Singer, W.C. Oliver and W.D. Nix, Met. Trans., Vol. 11A, p. 1895, 1980.

Appendix I. Processing of Al-Fe-Ni Alloys

Prealloyed air atomized powder of nominal weight composition Al-4.9%Fe-4.8%Ni was blended with pure Al powder and mechanically alloyed to give a final volume fraction of 0.19 Fe-Ni-Al_9. The powder was cold isostatically pressed in aluminum cans, outgassed, and hot extruded to full density. Three combinations of degassing and extrusion temperature were used in the range 370°C to 490°C, at a constant extrusion ratio of 16:1. The three combinations are designated: a (low degassing temperature, high extrusion temperature), b (low degassing temperature, low extrusion temperature), and c (high degassing temperature, low extrusion temperature).

Appendix II. Processing of Al-Ti Alloys

Prealloyed helium gas atomized powder of nominal weight composition Al-6%T was vacuum hot pressed and subsequently hot extruded to full density (Code AT6). A portion of the prealloyed powder was mechanically alloyed prior to vacuum hot pressing and hot extrusion (Code AM6).

In the third processing route, aluminum powder was mixed with Al_3Ti master alloy powder. The mix was then mechanically alloyed, vacuum hot pressed and hot extruded (Code MA6).

LIQUID PHASE SINTERING OF COMPOSITES

W.A.Kaysser

Max-Planck-Institut fuer Metallforschung
Institut fuer Werkstoffwissenschaften
Heisenbergstr.5, D-7000 Stuttgart 80, West Germany

Abstract

Pressureless sintering of composites often requires the presence of a liquid phase to eliminate all porosity. The shrinkage and the microstructural development during liquid phase sintering of almost all composites are governed by a number of basic principles, which will be outlined and illustrated by experimental results from sintering metal, ceramic, metal-ceramic and ceramic -glass systems. Major emphasis will be put on shrinkage phenomena, coarsening effects versus contact flattening, melt penetration effects and liquid film migration. Densification during liquid phase sintering is based on rearrangement and shape change of the solid constituent particles. To distinguish clearly between rate control by the mechanical movement produced by the capillary forces and rate control by dissolution-reprecipitation processes, these particle movements are described as primary and secondary rearrangements. It is demonstrated from the microstructural development of a number of samples that shrinkage is directly related to the shape change during grain growth. Grain boundary penetration by the liquid leads to skeleton disintegration ,but also to contact dilatation, particle dilatation and subsequently to particle disintegration. Important consequences of grain boundary penetration are boundary alloying and liquid film migration. It is shown that an appropriate heat treatment may induce liquid film migration and produce bulged grain boundaries which mechanically interlock the grains. Such a tailored microstructure improves the mechanical properties of Mo sintered with Ni considerably.

Processing and Properties for Powder Metallurgy Composites
Edited by P. Kumar, K. Vedula and A. Ritter
The Metallurgical Society, 1988

Introduction

Composites are multiphase materials where the microstructural combination and the component properties of the phases lead to novel or hybrid properties/materials. The necessary microstructure control in composites is oftern not possible with traditional forming processes like casting. Thus, powder metallurgy provides composition and microstructural control useful for forming composite materials. A classic example of a composite material is the hard metals with a very hard carbide phase embedded in a softer metal matrix /1/. More recent examples are aluminium alloys reinforced by SiC fibers or Si_3N_4 reinforced with SiC whiskers /2,3/. Powder metallurgy is a major technological route for the production of composite materials. Powders of the different phases are mixed and consolidated by pressureless sintering or by pressure assisted sintering (hot pressing, warm extrusion, hot isostatic pressing). Since most composites are high performance materials they ought to be free of pores after the sintering treatment. This goal is readily achieved by pressure assisted sintering /4/. In the more economic pressureless sintering often the presence of a liquid phase is required to eliminate all porosity /5/. The microstructural development and the shrinkage during sintering in the presence of a liquid phase are goverened by a number of basic principles, which apply to liquid phase sintering of almost all composites /6,7/. These principles will be outlined and illustrated by experimental results from sintering metal, ceramic, metal-ceramic and ceramic-glass systems. In each of these composites, one phase is solid at the sintering temperature and the other is the liquid. After processing the microstructure consists of two solid (oftern interpenetrating) networks. Major emphasis will be put on shrinkage phenomena, coarsening effects versus contact flattening, melt penetration effects and liquid film migration.

Shrinkage Phenomena

Densification during liquid phase sintering is based on rearrangement and shape change of the solid constituent particles (Fig.1) /8/. The driving force for both phenomena results from a decrease in energy connected to the changes in the areas of the liquid/vapour, liquid/solid and solid/vapour interfaces during rearrangement and/or shape change. Melt may originate from low melting point particles or areas which have built low melting point compositions by diffusion during heating. With the condition of good wetting, the liquid phase is pulled by capillary forces into particle necks and small pores. As a reaction to the capillary forces, particles are rearranged if their mobility allows it. This rearrangement of the solid particles is a necessary prerequisite for effective densification throughout the sintering process even during later sintering stages, when, for example the rates of particle shape change also limit the rearrangement rate. To distinguish clearly between rearrangement, which is essentially rate controlled by the mechanical movement produced by the capillary forces, and rearrangement, which is rate controlled by dissolution-reprecipitation processes, these particle movements are described as primary and secondary rearrangements. Shrinkage due to the shape change of the solid constituents always involves solution-reprecipitation processes and is therefore connected to secondary rearrangement in most cases.

A number of calculations show that high capillary forces are present when the amount of melt phase is small /9,10/. Pores which are much larger than the grains may exert no or small capillary forces if the wetting angle is >0 deg /11,12/. As a rule of thumb, the capillary pressure P in the melt resulting from a pore is estimated by $P = 2*\Gamma_{lv}/r_p$. Γ_{lv} is the specific energy of the liquid/vapour interface and r_p is the radius of the pore. If the melt is interconnected and at rest, then the capillary pressure is equal in the whole system /12/.

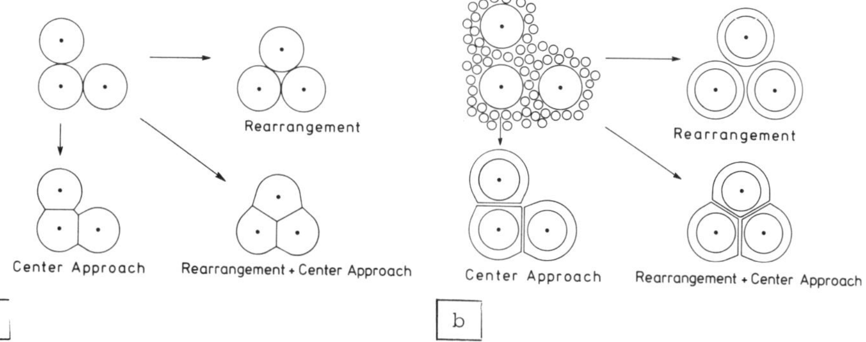

Figure 1 - Shrinkage by rearrangement and center approach (shape change) during liquid phase sintering. a) Monosized grains without coarsening. b) Grains with broad size distribution and coarsening.

Coarsening Effects versus Contact Flattening

Coarsening

Grain growth during liquid phase sintering of the composites determines their densification and their properties after processing. In many systems grain coarsening follows the relation

$$\bar{G}^3 = (\bar{G}^3 + kt)^{1/3} \qquad (1)$$

which indicates diffusion controlled Ostwald ripening as the rate controlling step /13,14/. Experiments with Fe-20wt.%Cu and Fe-30wt.%Cu samples also showed that during liquid phase sintering the particle size distribution, D(S), developed a steady state form which could be described by

$$D(S) = 2.136\ S^2 \exp(-0.712\ S^3), \qquad (2)$$

where 2S is the reduced particle size, $G/\bar{G}$, with $\bar{G}$ as the average particle size /14/.

When Price, Smithells and Williams published their famous micrographs of a W-Cu-Ni heavy metal alloy in 1937 they already recognized that grain growth and shape change of the solid constituents had occured simultaneously /15/. From recent experiments with mixtures of small tungsten powder (10μm) and large single crystal W seeds (220μm) that were liquid phase sintered in the presence of a Ni-rich melt, this idea was rejuvenated /16/. It was emphasized from the microstructural development of these samples that shrinkage is directly related to grain growth and shape accomodation. During liquid phase sintering the initial spherical shape of the single crystal seeds changed toward polyhedral shapes (Fig.2). Special etching showed the shape change of the spheres to be due only to the precipitated material /16/, that was supplied by dissolving smaller particles. Calculations of microstructural development showed similar general features as found in the

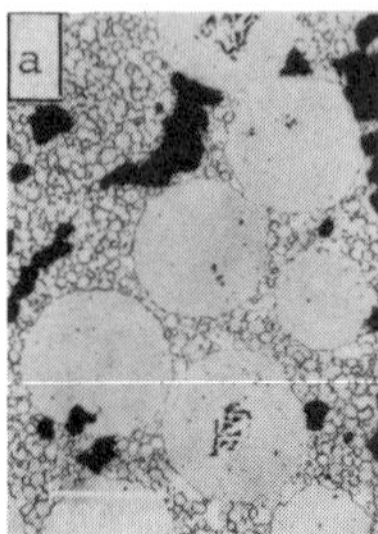

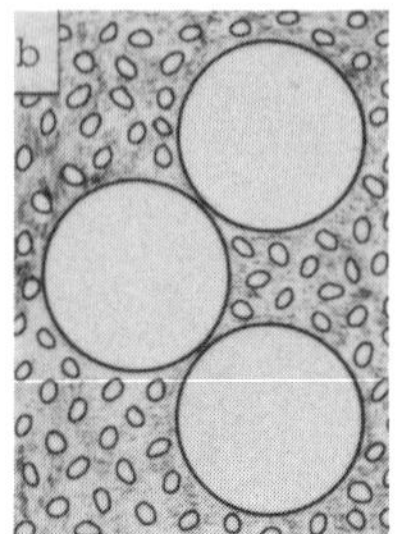

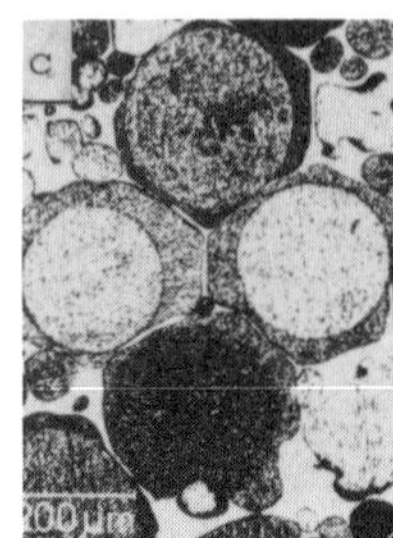

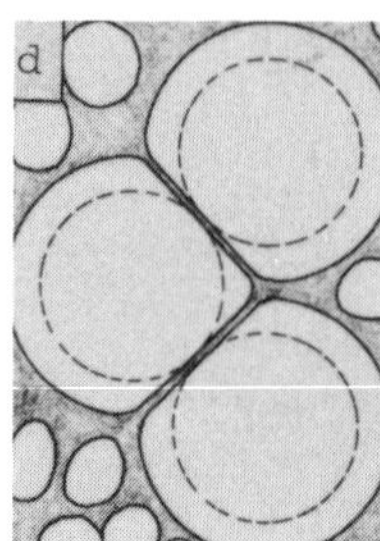

Figure 2 - Grain growth and shape accomodation during liquid phase sintering of W-Ni at 1670°C. a),c) - Microstructure after 3 and 120 min. b),d) - Calculated microstructure after 1 and 120 min.

liquid phase sintering experiments /17/. The calculations were conducted with the assumption that the large particles were taking part in the usual Ostwald ripening of the arrangement. No pressure in the flat contact areas was assumed beyond a value which kept the thickness of the thin liquid films in the contact areas between the particles at a constant value. As seen already from the experiments, the calculation showed that the deposition of material in the contact areas is extremely small in this system of large and small particles (Fig.2).

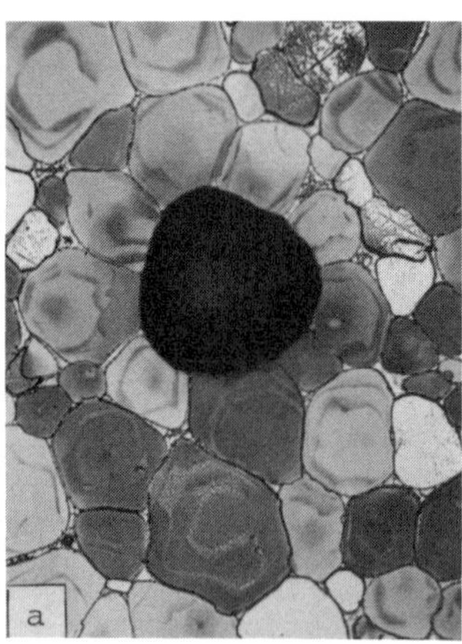

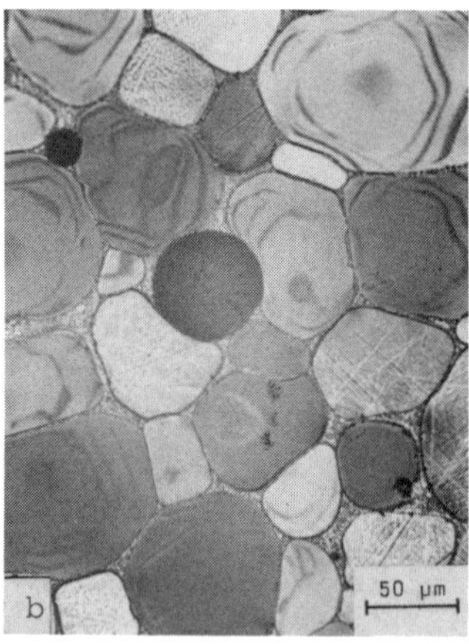

Figure 3 - Microstructure of Mo-4wt.%Ni after cyclic liquid phase sintering between 1300 and 1460°C. Held at 1460°C for 30 min in each cycle. Etched with Murakami's solution. a) - Microstructure around a pore /18/. b) - Microstructure around Al_2O_3 spheres deliberately added to the Mo-Ni powder mixture before sintering /19/.

During liquid phase sintering of a system with a steady state particle size distribution, similar shape accomodation was observed. During coarsening many grains rapidly accomodated their shape to a changing vicinity (Fig.3a) /18/. Even inert obstacles did not hinder further grain growth and were entrapped into growing grains /19/. This extreme of shape accomodation to an adjacent particle is shown in Fig.3b in Mo-Ni-Al_2O_3. The shape development of the Mo grains during annealing is indicated by etching boundaries. During liquid phase sintering with 2wt.%Ni at 1460°C the Mo grains trapped small Al_2O_3 spheres (black). No evidence of contact flattening was found.

Contact Flattening

Contact flattening is a classical approach to explain shape change and shrinkage during liquid phase sintering /20/. The contact flattening assumes thin liquid films in contact areas separating the particles. All particles transmit forces caused by the presence of pores. Close to the contact areas and through the thin liquid layers, the forces between the particles are thought to be transmitted whereby the stress is intensified in the contact areas compared to other areas of the particles. Thermodynamically, the increased pressure usually results in an increased solubility of the solid phase in the liquid. The concentration gradient due to the stress gradient in the liquid film and the adjacent solid areas yields a flow from the contact areas toward other lower stressed areas of the melt or the solid. In most cases the flow through the melt is assumed to be diffusion controlled. The continuous flow of dissolved material leaving the contact areas leads to a shape change of the particles and to the center approach of the particles.

Although many researchers have tried to find clear microstructural evidence for contact flattening as a major mechanism during liquid phase sintering, no unambiguous results have been published up to now. One of the major criticisms is that grain growth is completely neglected in the classical approach; although grain growth is much faster than the predicted contact flattening rates. A detailed calculation which took over all assumptions of the classical contact flattening theory but included in addition rapid grain growth showed that the contact flattening approach is incorrect /14/. The results of those calculations which included grain growth were in sharp contradiction to the variety of experimental results for those systems where shrinkage and grain growth during sintering had been measured.

Coarsening and Shrinkage

It must be concluded that shape accommodation during sintering occurs during grain coarsening and is due to small reprecipitation rate at particle contacts. Shrinkage may be vaguely described as secondary rearrangement which occurs when small particles are dissolved giving way for the movement of larger particles (Fig.1b). The model of contact flattening cannot describe the experimental observations of shape accommodation. If calculations based on the contact flattening concept, which include the observed rapid grain growth, were performed, then the results are in sharp contradiction with the experimentally measured shrinkage behavior.

Melt Penetration Effects

Grain boundary penetration by the liquid leads to skeleton disintegration ,but also to contact dilatation, particle dilatation and subsequently to particle disintegration. Important consequences of grain boundary penetration are boundary alloying and liquid film migration.

Skeleton Disintegration

An example of skeleton disintegration is shown in Fig.4. In many systems a rigid skeleton of small particles forms as a result of solid state sintering during heating to the temperature of liquid formation. When, in a Fe-30wt.%Cu material the Cu melts, it is instantaneously drawn into the small pores between the 2µm Fe-particles which form a rigid skeleton. No shrinkage or change of particle morphology is observed for times up to 2 min. This period ends with the collapse of particle chains as the melt penetrates along the grain boundaries, leading to rapid shrinkage of the sample by secondary rearrangement within the specimen /6/.

Grain Detachment and Particle Disintegration

Early grain boundary attack does not necessarily lead to densification. But if grain boundary attack is completed along all grain boundaries of a polycrystalline particle, particle disintegration and densification by secondary rearrangement may occur. This is illustrated in Fig.5a to c. In Al_2O_3-glass mixtures the shrinkage after particle disintegration due to rearrangement could be separated from shrinkage due to other primary rearrangement or solution-reprecipitation processes /6/. Up to 1400°C, the shape and size of the polycrystalline Al_2O_3 spheres was maintained and only primary rearrangement occurred (Fig.5b). Above this temperature grain boundary attack and particle disintegration occurred (Fig.5c) and secondary rearrangement led to considerable additional shrinkage (Fig.5a)/7,21/.

The attack of grain boundaries by a melt during sintering is most extensively studied in the Fe-Cu system in which compacts of Fe spheres with 10vol.% Cu spheres (Fig.6a), both of 100µm diameter were sintered at 1165°C for times up to 60 min /22/. After annealing for 3 min above the melting point of Cu, all contact regions between the Fe particles were filled with liquid (Fig.6b). After 8 min, grain boundary attack had led to severe changes in the contact area geometry; instead of flat interparticle contact areas, wavy layers of liquid were present and bumps consisting of Fe(Cu) solid-solution formed opposite each grain boundary . The additional melt in prior contact areas and at prior grain boundaries in the interior of the Fe particles caused contact and particle dilatation leading to macroscopic swelling /22/.

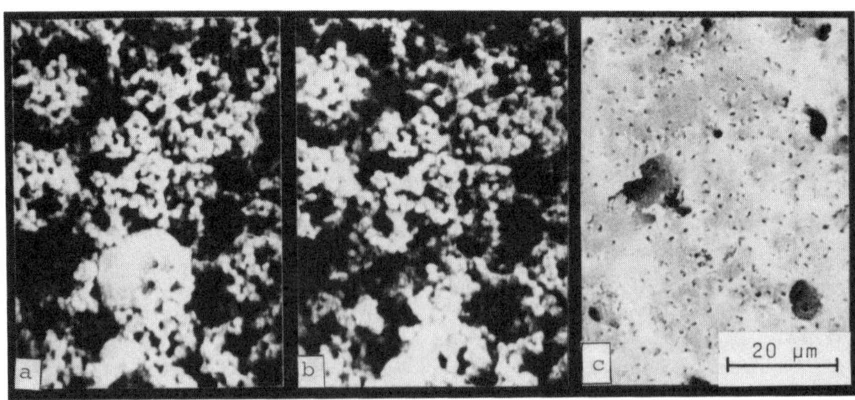

Figure 4 - Direct observation of skeleton disintegration during early stage of liquid phase sintering of carbonyl Fe (< 5µm) plus 30 wt.% Cu (10µm) at 1150 °C in a SEM with hot stage device /6/.

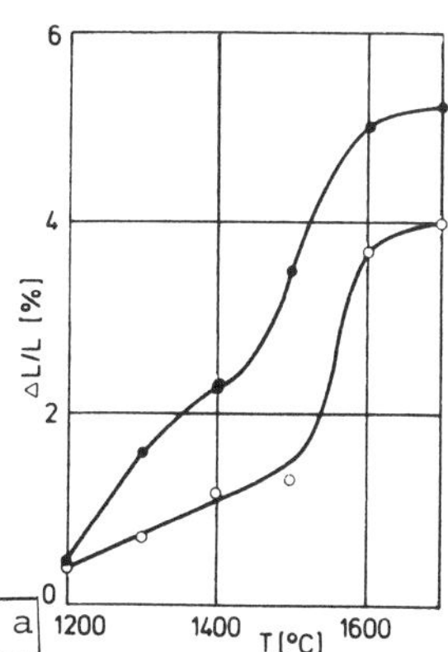

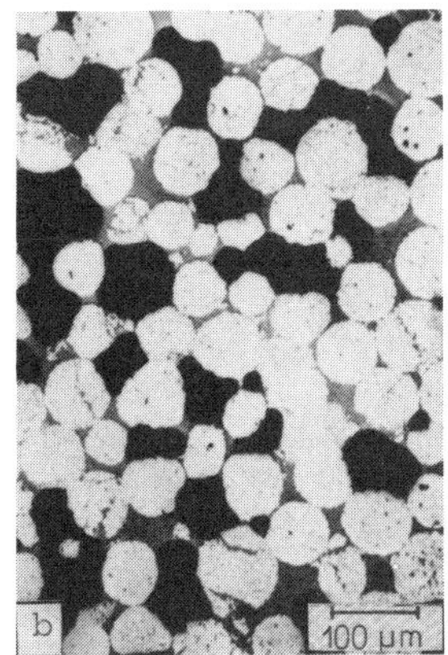

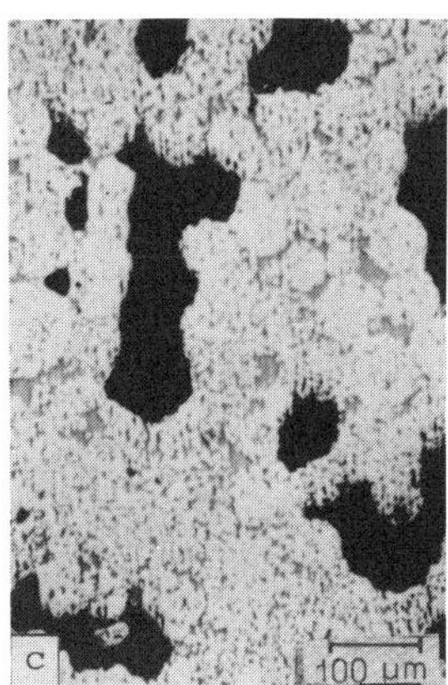

Figure 5 - Microstructure and shrinkage during liquid phase sintering of Al_2O_3-alkali borate glass /6,21/. a) - Shrinkage ● after 1 min, ○ after 120 min. b) - Microstructure after 1h at 1400°C. c) - Microstructure after 1h at 1700°C.

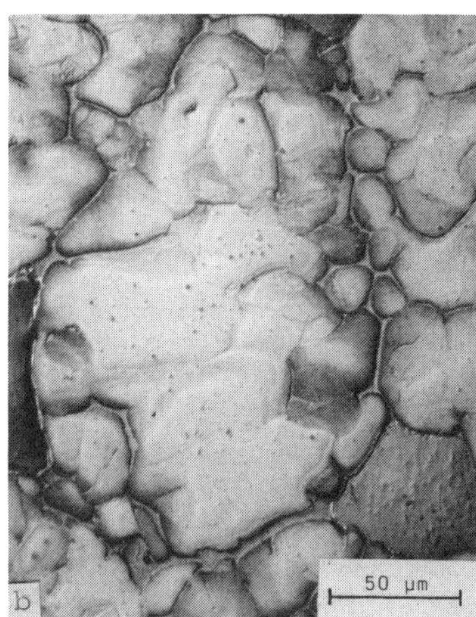

Figure 6 - Grain boundary penetration during liquid phase sintering of a Fe-10wt.%Cu compact at 1165°C /24/. a) - Before sintering. b) - After liquid phase sintering for 8 min.

Boundary Alloying

Melt penetration along grain boundaries results in preferential alloying of areas close to the prior grain boundaries. If the annealing time is short and bulk diffusion is slow, then unexpected phases may be present at the boundaries. These phases can determine the mechanical properties of liquid phase sintered composite materials as shown in the subsequent example. A Fe-10Cu-0.1C composition was mixed from spherical, polycrystalline Fe-0.011C (100μm) and from spherical Cu (100μm) powders. The compact (400MPa) was sintered in the presence of a liquid phase at 1110°C for 30s and cooled in the furnace. The white cementite at the boundaries in the interior of the Fe

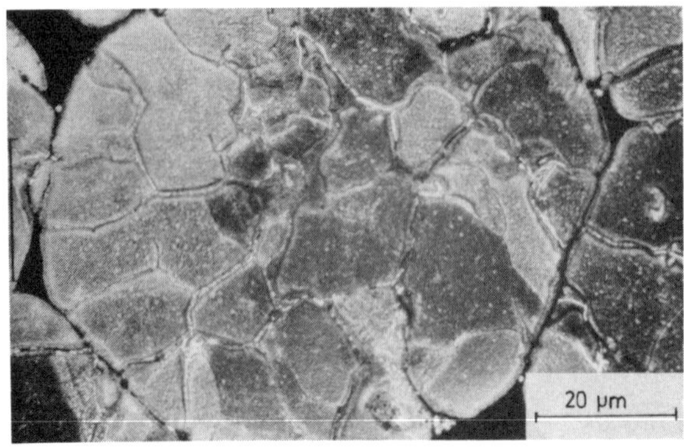

Figure 7 - Microstructure of Fe-10wt.%Cu-0.1wt.%C after liquid phase sintering at 1110°C for 30s /24/. The bright phase is cementite, the dark phase is Cu (etched with Klemm#3 solution /25/)

particle and the black Cu areas are clearly visible. During the short liquid phase sintering time Fe-Cu austenite has only formed in the intimate vicinity of the penetrated thin melt layers, whereas the regions in the interior of the grains remained Cu-free. Due to the fact that the formation of ferrite and perlite in Fe-Cu austenite occurs at lower temperatures than in Cu-free austenite /Lit/, formation of ferrite starts in the interior of the grains. The low solubility of carbon in ferrite results in an enrichment of carbon in the austenitic area of each grain which remains during growth of the ferrite phase. The last remaining austenite areas are the Fe-Cu solid solution layers in the vicinity of the boundaries. The enrichment of carbon in these layers results in a degenerated perlite formation in the boundary areas. It should be noted that the sample contains only 0.1wt.% carbon.

The penetration of melt along grain boundaries may also initiate liquid film migration. Liquid film migration is a more general phenomenon, however, which also occurs under much different conditions. Therefore liquid film migration will be described in a seperate section.

Liquid Film Migration

In recent years boundary migration in metals and ceramics, which is associated with a noticable concentration change in the wake of the moving boundaries, has been investigated intensively. It is kown as diffusion induced grain boundary migration (DIGM) /25/ and liquid film migration (LFM) /26,27/. The migrations are of considerable importance to solid state and liquid phase sintering of many composites.

When a well recrystallized Mo foil is annealed at 1380°C essentially no grain growth can be detected, but if a thin (un-reacted) Ni film is applied to the surface (or if the sample is exposed to Ni vapour), the grain boundaries develop bulges often looking like worms or snakes /28/ (Fig.8a). The length of the grain boundary and hence the grain boundary area is increased due to the formation of the bulges. Microprobe measurements show that the Ni concentration is increased in the boundaries, indicating a thin liquid film. Behind the migrating boundaries, a saturated solid solution is formed (Fig.8b). The Mo crystals in front of the migrating boundaries are completely free of Ni. From those measurements it became obvious that the formation of a solid solution may provide the driving force which allows the increase of the boundary area. Other measurements showed this type of boundary migration occurs when the solid solution left in the wake of the boundary has a lower free enthalpy than the material in front of the boundary /28/. Thus, the concentration in the wake of the migrating boundary may be higher or lower in solute concentration than in the front of the boundary. A second prerequisite is that the boundary provides a sink or source for the requested changes of the solute.

The free enthalpy change is necessary but does not yield itself the atomistic kinetics which cause boundary migration. The present atomistic model is based on coherency strains due to changes in the lattice parameters in the solid in front of the migrating boundaries /25 to 27/.

The solid in front of the migrating boundaries is not in equilibrium with the solute concentration of the boundary and solute will either deplete from or diffuse into this solid. When the boundary migrates at a constant velocity a steady state concentration profile will develop. The concentration gradients in the solid close to the liquid/solid interface will be very steep. If the lattice parameter changes with the concentration of the solute and coherency of the peripheral layers of the solid is maintained, coherency strains will arise. These coherency strains raise the free enthalpy of the solid which is in contact with the boundary. The solid in the wake of the migrating boundaries is thought to lose the coherency with the initial grain rapidly due to the incorporation of misfit dislocations.

This model was first proposed by Hillert /29/ and recently refined by Cahn et al. /26/. It explains nicely the effects of liquid film migration. Some modifications are required to explain the more complex stress states at grain boundaries during diffusion induced grain boundary migration.

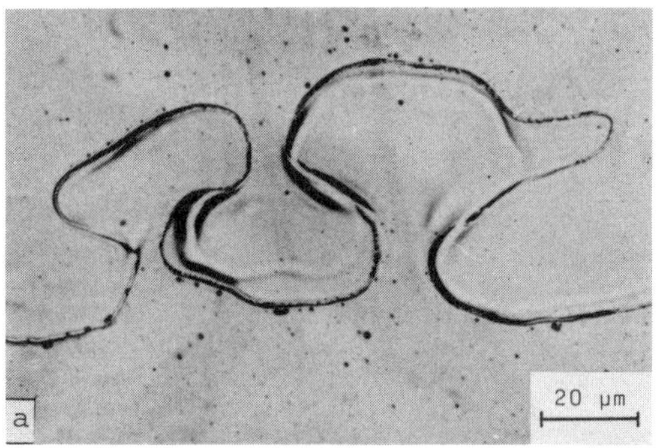

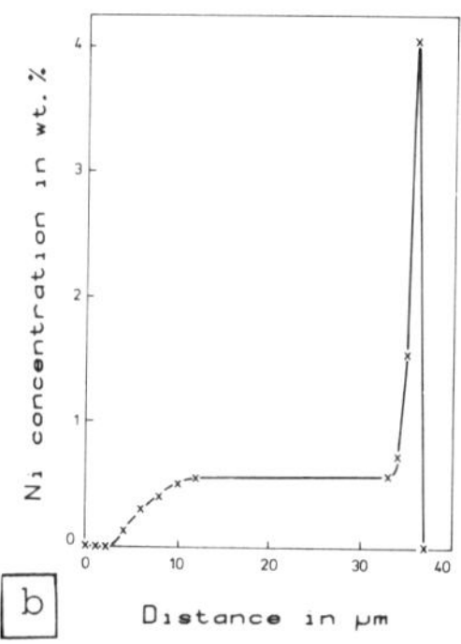

Figure 8 - Mo foil with a thin initially unreacted coating Ni layer after sintering at 1420°C for 10 min /28/. a) - Surface morphology. b) - Concentration profile across the migrated boundary shown in a).

The penetration of melt into contact areas and grain boundaries yields perfect conditions for LFM /2/. Figure 9 shows an area in a Fe particle after liquid phase sintering in the presence of Cu for 2 min. The thin liquid films have formed the bulges characteristic of LFM. Solid solutions of Fe-Cu have formed in the wake of the migrating boundaries /30/.

It must be emphasized that the bulk diffusivity of Cu in Fe is extremely small; thus, no solid solution formation is expected by bulk diffusion during the sintering times and temperatures applied usually in practice. The rapid homogenization is obtained by LFM, and hence LFM seems to be of great importance for this system.

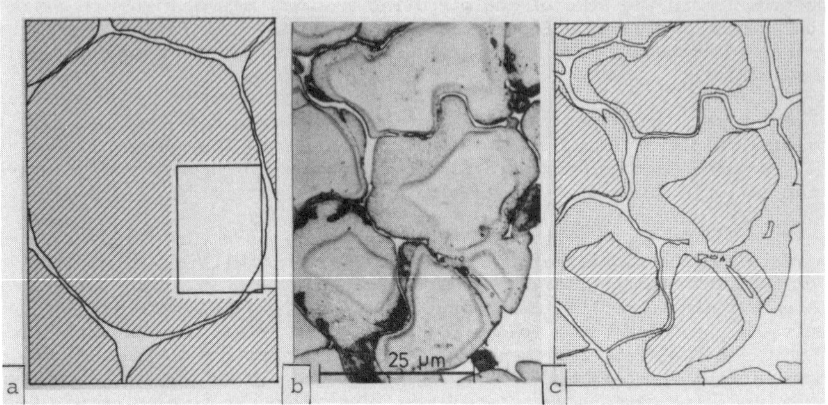

Figure 9 - Liquid film migration in Fe-10wt.%Cu during liquid phase sintering at 1120°C /30/. a) - Overview, schematic. b) - Microstructure. c) - Formation of solid solution, schematic; white: melt, dotted: solid solution Fe(Cu), lined: unalloyed Fe.

In Fe-Cu the penetration of melt along grain boundaries is suppressed if carbon is present in the alloy. However, penetration of contact areas is possible. Figure 10 shows the microstructure of Fe-10Cu-0.4C after liquid phase sintering for 20 min at 1130°C. The thin liquid layers in the contact areas are shifted by liquid film migration. In the wake of the moving boundaries a saturated solid solution of Fe-Cu is left, which mainly contains the pearlite which preferentially formed in these areas during cooling due to the fact that the eutectoid of Fe-C forms at a higher temperature than the eutectoid of Fe-Cu-C /6/.

Liquid film migration influences the mechanical properties of a number of systems. It is well known that W and Mo embrittle if sintered with activators like Ni. The Ni segregates at the grain boundaries. If an appropriate heat treatment is done, then LFM produces bulged grain boundaries (Fig.11a) which mechanically interlock the grains. This tailored microstructure im-

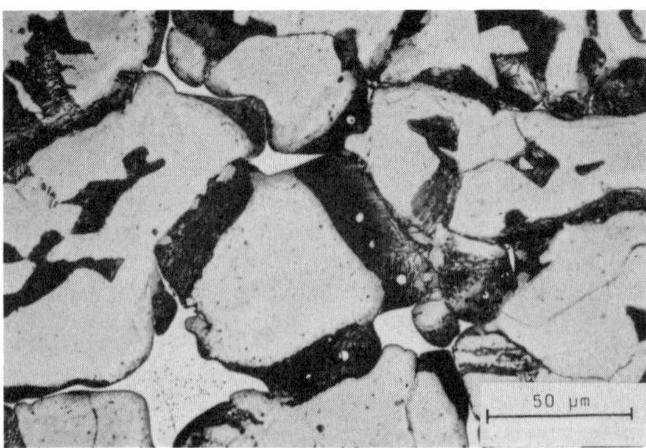

Figure 10 - Liquid film migration in Fe-10Cu-0.45C sintered at 1130°C for 20 min /6/.

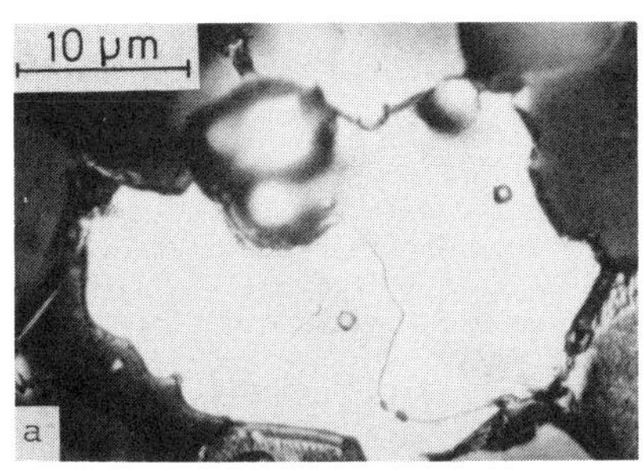

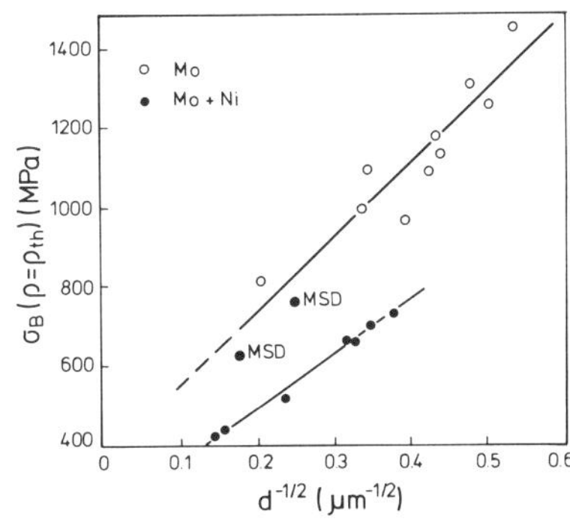

Figure 11 - Microstructure and bend strength of Mo-Ni after slow cooling. a) - Microstructure. b) - Bend strength, σ_B ; d is the average grain size, MSD cooling procedure resulted in bulged interlocking grain boundaries.

proves the mechanical properties considerably. Figure 11b shows the bend strength σ_B of Mo sintered as usual with Ni, and Mo sintered with Ni and heat treated to obtain LFM. Similar improvements might be possible in other systems susceptible to liquid film migration such as liquid phase sintered superalloys.

Achnowledgement

The continuous interest of Prof.G.Petzow is gratefully acknowledged. Parts of the work were supported by the Deutsche Forschungsgemeinschaft and the Internationales Buero of KFA Juelich. We thank Prof.R.M.German for fruitful discussions and his help with the final draft.

References

1. F.Benesovsky, "Carbide",Ullmanns Enzyklopaedie der technischen Chemie, vol.13 (Weinheim,FRG, Verlag Chemie, 1970), 13-30.
2. D.W.Petrasek at al., "Fiber-reinforced Superalloy Composites Provide an Added Performance Edge," Metals Progress, 130(1986)27-31.
3. G.C.Wei and P.F.Becher, "Development of SiC Whisker Reinforced Ceramics," Am. Ceram. Soc. Bull., 64(1985)298-304.
4. W.B.James, "New Shaping Methods for PM Components",Powder Metallurgy 1986 - State of the Art, eds. W.J.Huppmann, W.A.Kaysser and G.Petzow (Freiburg,FRG,Verlag Schmid,1986),71-100.
5. R.M.German, Liquid Phase Sintering (New York, Plenum Press, 1986).
6. W.A.Kaysser and G.Petzow,"Present State of Liquid Phase Sintering," Powder Metallurgy,, 28(1985)145-150.
7. G.Petzow and W.A.Kaysser, "Basic Mechanisms of Liquid Phase Sintering", Sintered Metal-Ceramic Composites,", ed. G.S.Upadhyaya, (Amsterdam, Elsevier Science Publishers, 1984),51-70.
8. G.Petzow and W.A.Kaysser,"Liquid Phase Sintering," Science of Ceramics 10, ed. H.Hausner (Weiden, FRG, Deutsche Keramische Gesellschaft, 1980),269-280.
9. W.J.Huppmann and H.Riegger, "Modelling of Rearrangement Processes in Liquid Phase Sintering,",Acta Metall., 23(1975)965-971.

10. K.S.Hwang, R.M.German and F.V.Lenel, "Capillary Forces between Spheres during Agglomeration and Liquid Phase Sintering,", Met. Trans., 18A(1984)11-17.
11. H.H.Park, O.J.Kwon and D.N.Yoon, "The Critical Grain Size for Liquid Flow into Pores during Liquid Phase Sintering," Met.Trans., 17A(1986)1915-1919.
12. W.A.Kaysser and G.Petzow, "Geometry Models for the Elimination of Pores during Liquid Phase Sintering in Systems with Incomplete Wetting," Science of Sintering, 16(1984)167-175.
13. W.A.Kaysser, S.Takajo and G.Petzow, "Particle Growth by Coalescence During Liquid Phase Sintering of Fe-Cu," Acta Metall., 32(1984)115-122.
14. W.A.Kaysser and G.Petzow, "Ostwald Ripening and Shrinkage during Liquid Phase Sintering," Z.Metallkde, 76(1985)687-692.
15. H.S.Price, C.J.Smithells and S.V.Williams, "Copper-Nickel-Tungsten Alloy Sintered with a Liquid Phase Present," J.Inst.Met., 62(1938)239-245.
16. D.N.Yoon and W.J.Huppmann, "Grain Growth and Densification during Liquid Phase Sintering of W-Ni," Acta Metall., 27(1979)693-698.
17. W.A.Kaysser et al., "Shape Accomodation during Liquid Phase Sintering," Contemporary Inorganic Materials: Progress in Ceramics, Metals and Composites, eds. G.Ondracek and O.Voehringer, (Juelich, Bilateral Seminars of the International Bureau KFA-Juelich CIM, 1986),217-222.
18. S.-J.L.Kang, W.A.Kaysser, G.Petzow and D.N.Yoon, "Liquid Phase Sintering of Mo-Ni Alloys for Elimination of Isolated Pores," Modern Developments in Powder Metallurgy, vol 15, eds. E.N.Aqua and C.M.Whitman (Princeton,N.J., MPIF-APMI, 1985),477-488.
19. S.-J.L.Kang, W.A.Kaysser, G.Petzow and D.N.Yoon, "Growth of Mo Grains around Al_2O_3 Particles during Liquid Phase Sintering," Acta Metall., 33(1985)1919-1926.
20. W.D.Kingery, "Densification during Sintering in the Presence of a Liquid Phase I," J.Appl.Phys., 30(1959)301-309.
21. S.Pejovnik, D.Kolar, W.J.Huppmann and G.Petzow, "Sintering of Al_2O_3 in Presence of Liquid Phase,"Sintering - Theory and Practice eds. D.Kolar, S.Pejovnik and M.M. Ristic (Amsterdam, Elsevier Scientific Publishers, 1982), 361-365.
22. W.A.Kaysser, W.J.Huppmann and G.Petzow, "Analysis of Dimensional Changes During Sintering of Fe-Cu," Powder Metallurgy, 23(1980)86-91.
23. Y.A.Chang et al., Phase Diagrams and Thermodynamic Properties of Ternary Copper-Metal Systems (New York, NY, International Copper research Organization, 1979),367-386.
24. W.A.Kaysser, "Boundary Alloying by Melt Penetration," to be published in Z.Metallkde..
25. G.Petzow, "Metallographisches Aetzen," Materialkundlich Technische Reihe 1 (Stuttgart, Gebrueder Borntraeger Verlag, 1976),67.
26. M.Hillert and G.Purdy, "Chemically Induced Grain Boundary Migration", Acta Metall., 26(1978)333-340.
27. D.N.Yoon, J.W.Cahn, C.A.Handwerker, J.E.Blendell and Y.J.Baik, "Coherncy Strain Induced Migration of Liquid Films through Solids," Interface Migration and Control of Microstructure (Washington, NBS, 1985).
28. C.A.Handwerker, J.W.Cahn, D.N.Yoon and J.E.Blendell, "The Effect of Coherency Strains on Alloy Formation: Migration of Liquid Films," Atomic Transport in Alloys: Recent Developments, eds. G.E.Murch M.A.Dayananda, (Dayton, OH, TMS/AIME Publications,1985).
29. M.Hofmann-Amtenbrink, W.A.Kaysser and G.Petzow, "Grain Boundary Migration in Recrystallized Mo Foils in the Presence of Ni," J. Physique, C4(1985),545.
30. M.Hillert, "On the Driving Force for Diffusion Induced Grain Boundary Migration",Scripta Metall., 17(1985)237-240.
31. W.A.Kaysser, "Liquid Film Migration During Sintering of Fe-Cu," to be pulbished in Z.Metallkde..

LIQUID PHASE SINTERING OF

PRESTRAINED HEAVY METAL ALLOYS

W.A.Kaysser and M.Yodogawa*

Max-Planck-Institut fuer Metallforschung
Institut fuer Werkstoffwissenschaften
Heisenbergstr.5, D-7000 Stuttgart 80, West Germany

* Now with Ryobi Limited, Tokyo, Japan

Abstract

A liquid phase sintered W-7Ni-3Fe composite was cold rolled to introduce strain in the single crystal grains. Liquid phase sintering after deformation resulted in thin liquid films penetrating into the boundaries in the neck areas, indicating an increased energy of these boundaries. Some of the thin liquid films moved out of the contact areas by stress induced boundary migration. During liquid phase sintering of heavily rolled material new small grains recrystallized in the original W grains. Thin melt layers penetrated along these boundaries and led to the disintegration of the original W grains. Disintegration reduced the average grain size by a factor of four.

Processing and Properties for Powder Metallurgy Composites
Edited by P. Kumar, K. Vedula and A. Ritter
The Metallurgical Society, 1988

Introduction

Research since 1975 has improved understanding of the major mechanisms leading to microstructural changes and to shrinkage during liquid phase sintering of heavy metal alloys /1 to 7/. However, only minor attention has been paid to the influence of increased dislocation densities, which may result from cold compaction of the powders prior to sintering or from cold deformation of the sintered composite materials. In most systems recrystallization occurs during heating to the liquid phase sintering temperature, hence only minor effects are expected during liquid phase sintering itself. In heavy metal alloys, where liquid phase sintering occurs before recrystallization starts, the increased dislocation density in some regions of the material may modify the grain growth behavior as indirectly suggested by previous work /8 to 14/. Clear explanations on the mechanisms were impossible because a broad variety of driving and retarding forces might have been effective during the experiments reported in the literature. A major cause for the broad variety of forces was the presence of pores and the presence of chemical non-equilibria. Pores hinder or retard the migration of grain boundaries or other interfaces which might be induced for example by the presence of high dislocation densities /15,16/. Chemical non-equilibria, in particular the non-equilibrium concentration of the W solid solution have been shown to induce liquid film migration (/6,13,17 to 20/).

In the present work liquid phase sintered 90W-7Ni-3Fe alloys were cold rolled to various reductions to introduce a variety of strains in the W particles. The deformed material was subsequently resintered. Chemical driving forces due to concentration changes of the liquid and solid phase during resintering were excluded by performing the initial sintering and the resintering treatment at the same temperature. In addition no pores were present during resintering of the prestrained W heavy metal alloys. Thus the microstructural development during resintering was essentially reduced to changes which were caused by the presence of an increased dislocation density.

Experimental Procedure

The 90W-7Ni-3Fe alloys were prepared from fine metal powders by liquid phase sintering at 1470°C in flowing dry hydrogen. The powders had mean particle sizes of 1.5μm (W), 5μm (Ni) and 3μm (Fe), and were obtained from H. C.Starck (Goslar, FRG), INCO (Düsseldorf, FRG) and BASF (Ludwigshafen, FRG), respectively. The final density of the sintered material was 18.4 g/cm^3 with no residual pores apparent in the sintered microstructure. This is slightly above the theoretical for the alloy due to matrix evaporation during sintering; thus the actual alloy composition is estimated as 92W-5.6Ni-2.4Fe. The sintered samples were cold rolled to give height reductions of 5, 15 and 50 %, without intermediate anneals. Subsequently the samples were liquid phase sintered at 1470°C. Another set of samples were sintered in the solid state region at 1420 °C after rolling. The samples were electrolytically polished and etched using Murakami's solution.

Results

After initial liquid phase sintering (prior to rolling) the alloy consisted of single crystal W(Fe,Ni) grains (subsequently called original W grains) embedded in a Fe- and Ni-rich matrix. The W grains were connected at the contact regions by boundaries. Figure 1 shows the microstructure of samples which were annealed in solid state at 1420 °C for 3 min after cold rolling. In the sample reduced by 15 %, new grains with diameters of approximately 6 μm recrystallized at a few contact areas between the original W grains (arrows in Fig.1a). In samples reduced by 50 %, several new grains formed at most contact areas between the original W grains (Fig.1b). The

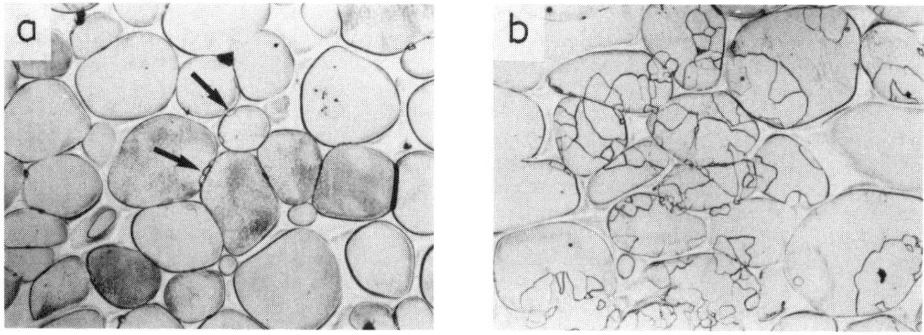

Figure 1 - Microstructures of a prestrained W-7Ni-3Fe heavy metal alloy solid state sintered at 1420°C for 3 min. a) - 15 % reduction. b) - 50 % reduction.

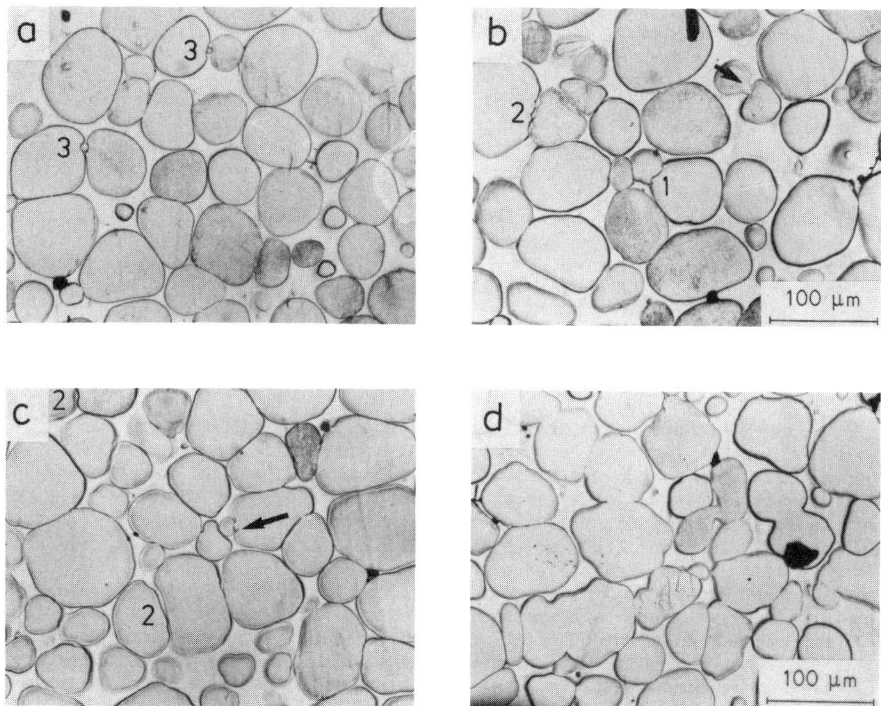

Figure 2 - Microstructures after liquid phase sintering of a 5 % prestrained W-7Ni-3Fe heavy metal alloy at 1470 °C for a) - 1 min, b) - 10 min, c) - 40 min and d) - 720 min.

microstructures of prestrained samples after liquid phase sintering at 1470°C are shown in Figs.2, 3 and 4. After short liquid phase sintering of the samples, six distinct microstructural features were found, which had not been present in the microstructures of the material after initial liquid phase sintering. i) New grains with diameters between 5 and 35 μm showed up at the necks and in the interior of original W grains (Fig.4a, indicated by 3 in Fig.2a). ii) Original W grains grew at the expense of their immediate neighbor despite the simultaneous increase in interface area caused by this process (indicated by arrows in Fig.2b and c). iii) Thin liquid films penetrated along the grain boundaries and triple junctions of the newly formed grains, leading to detachment and to disintegration of the original W grains (Fig.4b). iv) Thin liquid films were present in many contact areas replacing the initial boundaries (Fig.3a). v) Some thin liquid layers between the original W grains developed a wavy shape (indicated by 2 in Fig.2b and c). vi) Grain boundaries in the necks between original W grains developed a wavy shape (indicated by 1 in Fig.2b).

After a short liquid phase sintering treatment (up to 10 min) of samples which were reduced by 5%, the formation of new grains was restricted to the immediate contact region. The diameters of the new grains were between 5 and 7 μm. In addition the features iv),v) and vi) were found to a large extent. During prolonged sintering (40 min and more) the features ii) became more characteristic (Fig.2c). In 15% reduced samples liquid phase sintering at 1470 °C for 1 and 10 min led to the formation of new grains in the contact areas between the particles with average diameters of 5 and 10 μm (Fig.3a and b). Some of the newly formed grains still appear to grow into the interior of the single crystal grains after prolonged liquid phase sintering (arrows in Fig.3c). After liquid phase sintering for 720 min the boundaries between the particles were mostly straight again (Fig.3d).

As mentioned above the majority of initial boundaries in the neck areas are replaced by thin liquid layers during liquid phase sintering of the prestrained samples. The particle contact area ratio, R_1, is defined as the ratio of contact area where two particles are separated by a thin liquid film to the particle contact area where the particles are separated by a boundary of the type present after initial liquid phase sintering. The changes in R_1 with sintering time at 1470 °C following 5 and 15% deformation are shown in Fig.5. The replacement of grain boundaries by liquid films is more pronounced in alloys reduced by 5 % than in the alloys which are reduced by 15 % during cold rolling.

Figure 4 shows the microstructures of the cold rolled (reduction 50 %) W-7Ni-3Fe heavy metal alloy after liquid phase sintering at 1470°C. After annealing 1 min many new grains with diameters smaller 5 μm have formed at most contacts (Fig.4a). Larger grains formed in the interior of the original W grains (Fig.4b). After liquid phase sintering for 10 min only a few areas in the particles were left, where no new grains formed (triangles in Fig.4b). The formation of new grains is followed by the penetration of melt along the newly formed grain boundaries. At the periphery of the original W grains the new grains are detached and appear to move toward the melt-rich pools. After 40 min of annealing disintegration of the original W grains is pronounced (Fig.4c). Small newly formed and detached grains visible in the microstructure after sintering 10 min partly disappeared or spheroidized in the annealing interval between 10 and 40 min (compare Fig.4b and c). After prolonged liquid phase sintering (720 min, Fig.4d) the microstructure shows the usual coarsened grains, very similar to the initial microstructure of the liquid phase sintered undeformed samples.

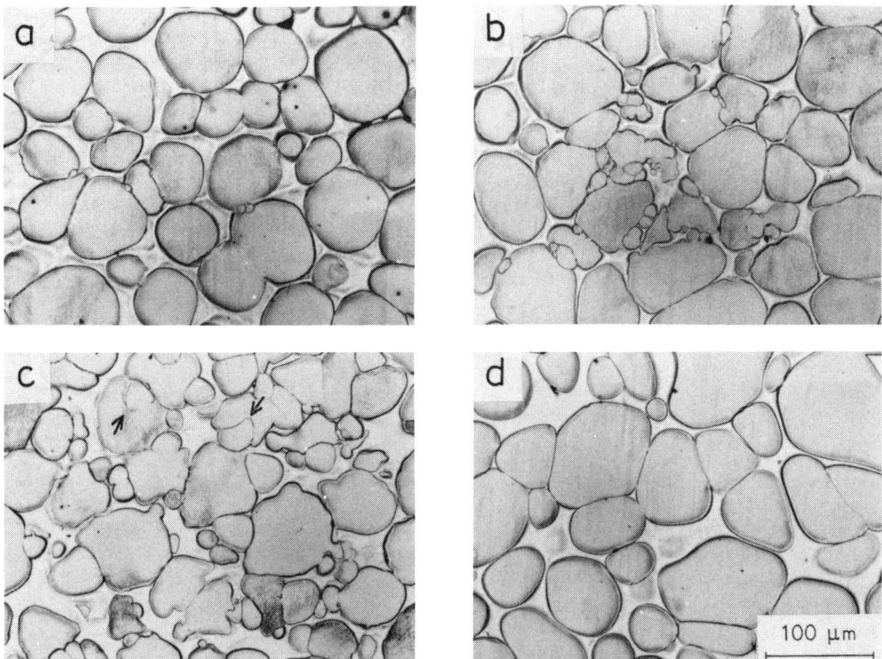

Figure 3 - Microstructures after liquid phase sintering of a 15 % prestrained W-7Ni-3Fe heavy metal alloy at 1470 °C for a) - 1 min, b) - 10 min, c) - 40 min, d) - 720 min.

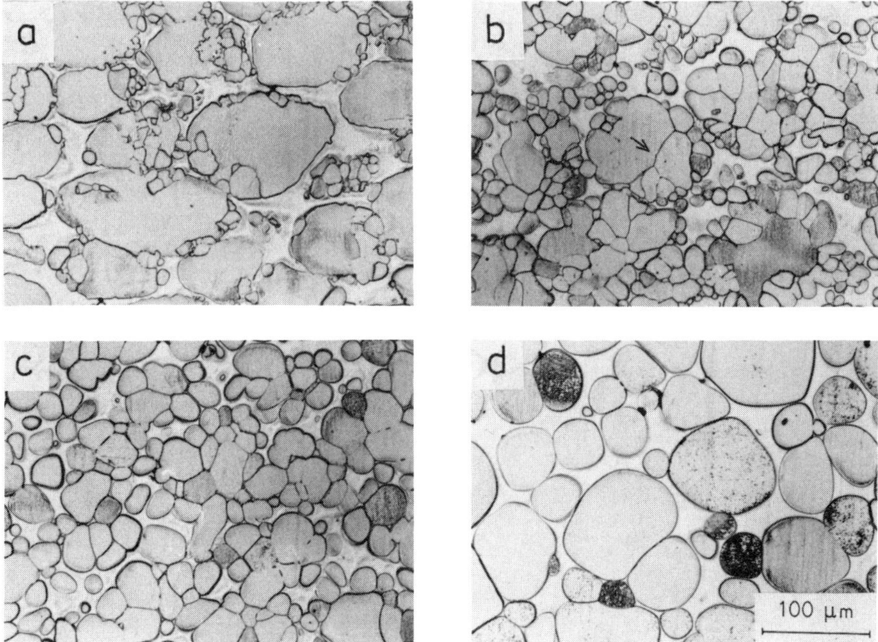

Figure 4 - Microstructures after liquid phase sintering of a 50 % prestrained W-7Ni-3Fe heavy metal alloy at 1470 °C for a) - 1 min, b) - 10 min, c) - 40 min, d) - 720 min.

Figure 6 shows the average chord length L_3 of the W(Ni,Fe) particles and the detached grains after sintering samples which had been reduced 50 % by cold rolling. The state after solid state sintering at 1420°C for 3 min is indicated by the asterisk. During liquid phase sintering the formation of new grains and the particle disintegration result in a drastic grain refinement which is still present after annealing for 40 min. During prolonged liquid phase sintering the grains coarsen again. The average intercept length, L_3, measured at times over 40 min fits the relation $L_3(t)^n = L_3(2400)^n + k*t_s$, with $n = 1/3$, $k = 3.41*10^{-18}\ m^3s^{-1}$ and $t_s = t - 2400$ s, where t is the isothermal annealing time.

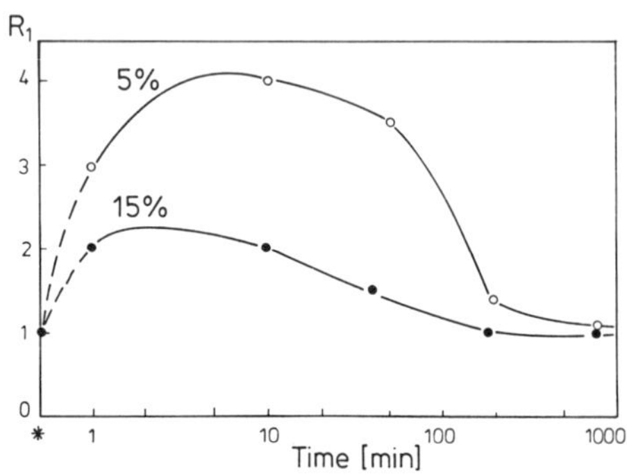

Figure 5 - Particle contact area ratio, R_1, in cold rolled W-7Ni-3Fe after liquid phase sintering at 1470°C (* after solid state sintering at 1420 °C for 3 min). R_1 is defined as the ratio of contact area where two particles are separated by a thin liquid film to the area where the particles are separated by a boundary of the type present after initial liquid phase sintering.

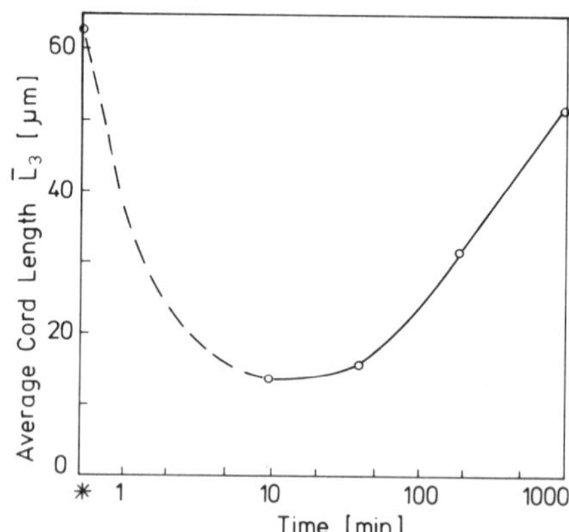

Figure 6 - Average chord length of grains of a W-7Ni-3Fe heavy metal alloy, reduced 50 % by cold rolling, after liquid phase sintering at 1470 °C (* indicates the state after solid state sintering of the prestrained sample at 1420 °C for 20 min).

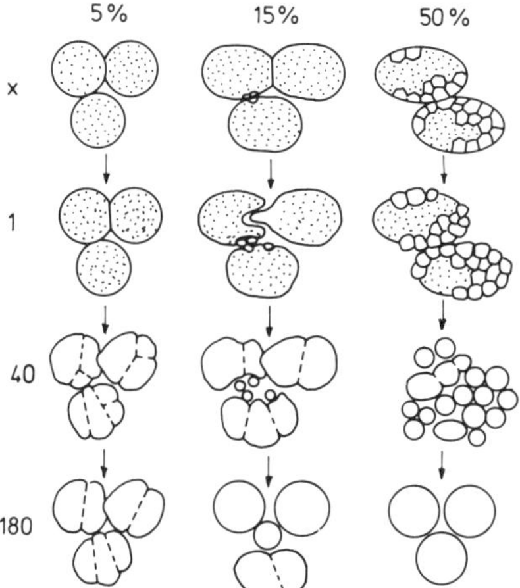

Figure 7 - Schematic description of the microstructural development during solid state and liquid phase sintering of a prestrained W-7Ni-3Fe alloy. x after solid state sintering at 1420°C for 3 min; 1, 40, 180 after liquid phase sintering at 1470°C for 1, 40 and 180 min.

Figure 7 shows a schematic summary of the microstructural development observed during solid state and liquid phase sintering of the prestrained W-7Ni-3Fe alloy.

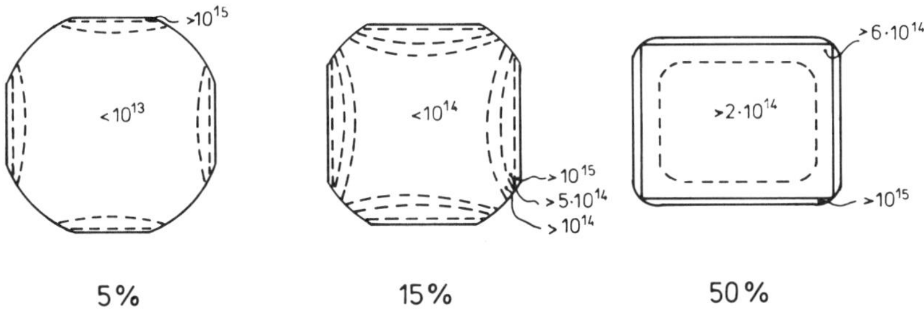

Figure 8 - Dislocation density (in m^{-2}) in W grains of prestrained W-7Ni-3Fe. Reduction by cold rolling in %.

Discussion

Recrystallization, Grain Growth and Dislocation Density

During solid state and liquid phase annealing of cold rolled W heavy metal alloy new grains form in contact areas and in the interior of the original single crystal W(Ni,Fe) grains. The formation and growth of the new grains depends on the energy reduction through the elimination or rearrangement of dislocations. The reduction in strain energy must overcome the energy increase which results from the formation of additional interface area when the new grains form and grow. The present observations will be discussed at first with respect to the magnitude of both energies, as well as the resulting driving forces.

The maximum potential difference at a grain boundary due to the dislocation density difference at either side is obtained when the coherency of the adjacent grains is completely lost and the energy of the dislocations transforms directly without relaxation into an average increase of the chemical potential, $\Delta\mu_1$. For this case the increase in the chemical potential is estimated by /21,22/

$$\Delta\mu_1 = \Omega E_v \Delta\rho \qquad , \qquad (1)$$

where E_v is the average specific line energy of the dislocations and Ω is the molar volume of the host material and $\Delta\rho$ is the change in the dislocation density. The isolated approximately spherical grains in the originally single crystal W grains tend to shrink due to their curved interfaces. The unrelaxed potential difference at either side of the grain boundary due to the curvature is /22/

$$\Delta\mu_2 = -4\gamma_{gb}\Omega / G \qquad , \qquad (2)$$

where 4/G is the sum of the two principal curvatures. In approximation G is the grain size, and γ_{gb} is the grain boundary energy. Calculations with $\Delta\mu_1 = \Delta\mu_2$ indicate that the reduction of dislocation density by the newly formed grains has to be considerable to overcome the surface tension of the curved interface.

After liquid phase sintering of 5% reduced samples new grains were observed in the contact areas with sizes between 5 and 7 µm, which is equivalent to a reduction in dislocation densities in the order of 10^{15} m^{-2}. Larger grains were not observed, which indicates that the deformation outside the direct contact areas is small and that the dislocation density decreases steeply to much lower values within 3 µm from the contact plane.

After sintering samples which were reduced 15 % at 1420°C for 3 min the diameters of the newly formed grains (Fig.1a) are less than 6 µm, hence the formation of the grains required a minimum dislocation density greater than 10^{15} m^{-2}. The deformation of contact areas yields dislocation densities on this order of magnitude and higher, within 3 µm from the contact planes. During liquid phase sintering the decrease of the dislocation density in the wake of moving boundaries caused thin liquid films to migrate from the contact areas against their center of curvature into one of the adjacent grains indicating a dislocation density of $5*10^{14}$ and $1.4*10^{14}$ m^{-2} at distances of 15 and 25µm from the contact areas, respectively.

After liquid phase sintering of 50% reduced samples, grains smaller than 5 and 9µm in size were present in the contact areas and in other peripheral areas of the original W grains, corresponding to equivalent dislocation densities above $2*10^{15}$ and $6*10^{14}$ m^{-2}, respectively. In the interior of the original W grains dislocation densities of $1.3*10^{14}$ m^{-2} were indicated. The microstructures in Fig. 4a and 4b indicate that the recrystallization starts at the periphery of the particles and proceeds toward the interior of the particles. The delayed start of recrystallization and the larger size of the newly formed grains in the inner parts of the particles are likely to be due to the lower dislocation densities in the interior of the particles.

Figure 8 shows schematically the typical minimum dislocation densities in various areas of the prestrained original W grains obtained from the size of grains newly formed during solid state and liquid phase sintering of the prestrained material.

<u>Liquid Film Penetration during Liquid Phase Sintering</u>

The penetration mechanism of thin liquid films along the boundaries between original W grains and along the grain boundaries of newly formed grains is not yet clear, despite the important changes of the microstructure caused by the penetration, e.g. a reduction of the grain size instead of the usual grain coarsening.

After liquid phase sintering of the slightly deformed alloy (5% reduction) thin liquid films (thickness < 1 μm) became visible in many contact areas. This means the initial boundaries were replaced by liquid phase layers (Figs.2a, 2b and 4). It requires that the specific boundary energy γ_{gb} is smaller than twice the surface energy γ_{sl} before deformation and that γ_{gb} is larger than 2 γ_{sl} after deformation. It is suggested that the excess free energy of the boundary in the presence of high dislocation densities near to the boundaries, (that means after deformation) composes of the increased energy of the atoms at the interface itself and of the increased energy of thin adjacent bulk layers.

Figure 9 shows a model where a liquid film of approximately 1μm thickness penetrates along prior grain boundaries /23/. At the top of the penetrating liquid film, material with high dislocation density dissolves in the melt, diffuses along the melt layer and precipitates at solid/liquid interfaces which are not part of the contact areas /24/. With an initial thickness of the "liquid film" of 1.24 10^{-10} m (roughly a monolayer of Ni being segregated at the boundary) the model yields 1016, 164 and 32 s for the complete penetration into a boundary at the contact area with the dislocation densities of 10^{13}, 10^{14} and 10^{15} m^{-2}. These values fit well to the experimentally observed times (compare Figs. 2, 3 and 4). In addition some evidence of the penetration as described is given from the microstructures (x in Fig.3).

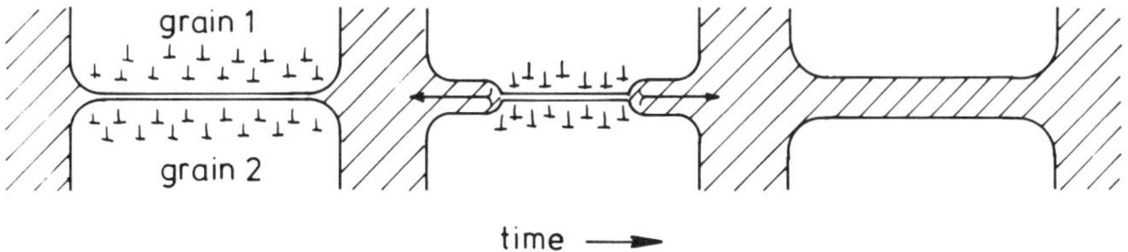

Figure 9 - Schematic representation of the penetration of a melt layer along a deformed boundary between original W grains.

Disintegration of Original W Grains

During prolonged liquid phase sintering grains newly formed in original W grains are detached from the original grains and move toward adjacent liquid pools (Fig.4b). The detachment of grains from the polycrystalline areas requires the penetration of melt along the grain boundaries. The detachment also requires sufficient cross sectional area for the melt flow during the grain movement and a certain spheroidization of the grains at triple junctions and at the nodes of four adjacent grains. The spheroidization is equivalent to the opening of triangular melt channels along the triple junctions of grain boundaries. The melt channels at triple junctions of grain boundaries tend to increase their size if γ_{gb} is larger than 3 γ_{sl}. The penetration of melt into grain boundaries between two grains requires $_{gb}$ to be larger than 2 γ_{sl}. Calculations on various models suggest that a fast opening of the melt channels along triple junctions depends on a prior melt penetration along the planar boundary areas /25/.

Conclusions

In samples which are 5% reduced the deformation is essentially restricted to the immediate contact areas. Due to thin dislocation-rich layers at the boundaries, most of the initial boundaries between the particles were replaced by thin liquid films of up to 1μm. The penetration occurs by a mechanism where material of increased dislocation density dissolves at the front of the penetrating liquid film and reprecipitates with lower dislocation density at other grains.

After 15% reduction the resintered samples show liquid film migration driven by the difference in dislocation density of the dissolving and the reprecipitating material. The dislocation densities are 10^{15} m^{-2} in the contact areas with a depth of up to 10 μm and between $1*10^{14}$ and $8*10^{14}$ m^{-2} in the interior of the original W grains.

50% reduction of the original W grains yields an excess dislocation density of $5*10^{15}$ m^{-2} at the periphery of the grains which decreases continuously to $5*10^{13}$ m^{-2} toward the interior of the original grains. The boundaries of grains forming throughout the original W grains during liquid phase sintering are subsequently penetrated by thin liquid films. Melt channels at triple junctions widen until detachment of the grains occurs, resulting in a complete disintegration of the recrystallized original grains. This mechanism reduces the average grain size by a factor of four.

Acknowledgement

The authors thank Prof.G.Petzow for his continuous interest. The work was partly supported by the Deutsche Forschungsgemeinschaft. The fruitful discussions with Prof.R.M.German on the final draft are acknowledged.

References

1. W.J.Huppmann et al., "The Elementary Mechanisms of Liquid Phase Sintering - I. Rearrangement," Z.Metallkde., 70(1979)707-713.
2. W.J.Huppmann, "The Elementary Mechanisms of Liquid Phase Sintering - II. Solution-Reprecipitation," Z.Metallkde., 70(1979)792-797.
3. G.Petzow and W.A.Kaysser, "Basic Mechanisms of Liquid Phase Sintering," Sintered Metal-Ceramic Composites, ed. G.S.Upadhyaya (Amsterdam, Elsevier Science Publishers, 1984),51-70.
4. W.A.Kaysser and G.Petzow, "Present State of Liquid Phase Sintering," Powder Metallurgy, 28(1985)145-150.
5. D.N.Yoon and W.J.Huppmann, "Grain Growth and Densification during Liquid

Phase Sintering of W-Ni," Acta Metall. 27(1979)693-698.
6. D.N.Yoon and W.J.Huppmann, "Chemically Driven Growth of Tungsten Grains during Sintering with Liquid Nickel," Acta Metall. ,27(1979)973-977.
7. S.S.Kim and D.N.Yoon, "Coarsening Behaviour of Mo Grains Dispersed in Liquid Matrix," Acta Metall., 31(1983)1151-1157.
8. E.G.Zukas and D.T.Eash, "Possible Reinforcement of the W-Ni-Fe-Composite with W Fibers," J.Less-common Met., 32(1973)345-353.
9. E.G.Zukas, P.S.Z.Rogers and R.S.Rogers, "Unusual Spheroid Behaviour During Liquid Phase Sintering," Int.J. of Powd.Met.and Powd. Tech., 13(1977)27-33.
10. G.Petzow and W.J.Huppmann, "Fluessigphasensintern," Z.Metallkde., 67,(1976)576-582.
11. M.Hofmann-Amtenbrink et al., "Kosseluntersuchungen zur Korngrenzenwanderung in verformtem Wolfram in Anwesenheit von Nickel," Z.Metallkde., 77(1986)368-376.
12. W.A.Kaysser, F.Puckert and G.Petzow, "Recrystallization and Grain Boundary Migration During Sintering," Powder Metallurgy International, 12(1980)188-191.
13. W.A.Kaysser and S.Pejovnik, "Grain Boundary Migration During Sintering of Mo with Ni Additions," Z.Metallkde., 71(1980)649-653.
14. W.Schatt et al., "Influence of Dislocations on Solid State and Liquid Phase Sintering," Powder Metallurgy Int., 19(1987)14-19, and 19(1987)37-39.
15. R.Brook, "Pore-grain Boundary Interactions and Grain Growth," J.Am. Ceram.Soc., 52(1969)56-57.
16. C.H.Hsueh,A.G.Evans and R.L.Coble, "Microstructure Development during Final/Intermediate Stage Sintering - I. Pore/Grain Boundary Separation," Acta Metall., 30(1982)1269-1279.
17. C.A.Handwerker, J.W.Cahn, D.N.Yoon and J.E.Blendell, "The effect of Coherency Strains on Alloy Formation: Migration of Liquid Films," Atomic Transport in Alloys: Recent Developments", G.E.Murch ed. M.A.Dayananda (Dayto, OH, TMS/AIME Publications, 1985).
18. D.N.Yoon, J.W.Cahn, C.A.Handwerker, J.E.Blendell and Y.J.Baik, "Coherncy Strain Induced Migration of Liquid Films through Solids", Interface Migration and Control of Microstructure (Washington D.C., NBS, 1985).
19. W.-H.Rhee, Y.-D.Song and D.N.Yoon, "A Critical Test for the Coherency Strain Effect on Liquid Film and Grain Boundary Migration in Mo-Ni-(Co-Sn) Alloy," Acta Metall., 35(1987)57-60.
20. M.Hofmann-Amtenbrink, W.A.Kaysser and G.Petzow, "Grain Boundary Migration in Recrystallized Mo Foils in the Presence of Ni," J. Physique, C4(1985)545-552.
21. P.A.Beck, P.R.Sperry and H.Hu, "The Orientation Dependence of the Rate of Grain Boundary Migration,",J.Appl.Phys., 5(1950)420-425.
22. H.P.Stuewe, "Driving and Dragging Forces in Recrystallization", Recrystallization of Metallic Materials, ed. F.Haessner (Stuttgart, Riederer Verlag, 1978),11-21.
23. W.A.Kaysser, W.J.Huppmann and G.Petzow, "Analysis of Dimensional Changes During Sintering of Fe-Cu," Powder Metallurgy, 23(1980)86-91.
24. S.-J.L.Kang et al, "Growth of Mo Grains around Al_2O_3 Particles During Liquid Phase Sintering," Acta Metall., 33(1985)1919-1926.
25. W.A.Kaysser, "Influence of Dislocations on the Basic Mechanisms of Liquid Phase Sintering in W-Ni-Fe," to be submitted to Int.J.Refractory and Hardmetals.

COMPOSITE STRUCTURES PRODUCED

BY LOW PRESSURE PLASMA DEPOSITION

MR Jackson, PA Siemers, JR Rairden, RL Mehan and AM Ritter

General Electric Company
Corporate Research and Development
Schenectady, NY 12301

Abstract

Powder particles injected into a plasma are melted, accelerated in the plasma jet, and then impacted and solidified against a substrate or target in a low pressure chamber. In this way, deposits are produced with thicknesses in the range of 100 μm to 5 cm. The deposits are collections of "pancake-shaped" solidified disks, each several microns in thickness. For monolithic materials, theoretical density is nearly achieved (97-99+%) for metals with melting points ranging from that of Al to that of Mo. This technique is well suited to fabrication of composite structures. Since the starting materials are powders, sequential feeding of different materials into the plasma will produce a laminated structure. Depending on feed rate and length of spray time for each material, the laminate spacing and volume fractions of the phases can be controlled. The spacing and/or volume fractions can be varied throughout the structure, again by controlling the particles that are fed into the plasma.

The process has been used to produce a number of different composite structures, including metal-metal, metal-oxide, and metal-carbide systems. With a view toward high temperature applications, the matrix compositions receiving the greater attention are iron or nickel-base alloys. Both physical behavior (elastic modulus and thermal expansion) and mechanical behavior (tensile and rupture strength) have been characterized for such materials.

Processing and Properties for Powder Metallurgy Composites
Edited by P. Kumar, K. Vedula and A. Ritter
The Metallurgical Society, 1988

Introduction

Designs for advanced aircraft propulsion systems are aimed at doubling the thrust-to-weight performance of current engines (1). High temperature composites are prime candidates to achieve the much higher specific strength and stiffness needed to reach these design goals. Production of such composites can be extremely expensive, especially when complex shapes are required. Low pressure plasma deposition (LPPD) is a simple process that can be used to produce relatively inexpensive net shape or near-net shape components made of high temperature composite materials (2,3). In this paper, the process, the resulting structures, and composite physical and mechanical behavior for several high temperature systems will be described. The discussion will be limited to structures where both phases of the composite pass through the plasma gun.

The Process

Arc plasma guns are commonly used to deposit coatings on metal components in air and inert environments (4,5). Free-standing bodies also can be produced by such spray casting methods (6,7). Incorporation of the arc gun into a low pressure chamber has resulted in coatings of lower porosity and much lower oxygen contamination levels (8,9). Extension of the LPPD process from coatings to structures has led to deposition of thick structures several cm in thickness to produce rapidly solidified, dense, near-net shape structures (1,10) of Ni-base superalloy materials.

In the LPPD process, shown schematically in Figure 1, a mandrel to serve as a deposition substrate is placed in a chamber. The mandrel may be incorporated into the final component, or it may be sacrificial, in which case it will later be separated from the deposit chemically or mechanically. The chamber is typically cylindrical, 1.5-2m in diameter, and 1-2m in length, and is evacuated with a mechanical pumping system. After chamber pressure reaches a sufficiently low value, <10 torr, the plasma gun is ignited. In the dc

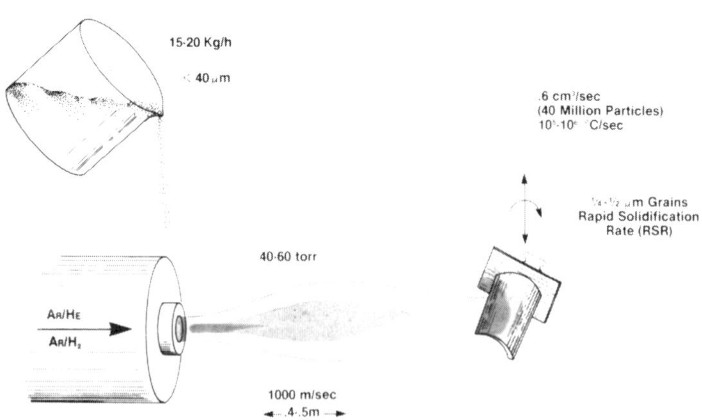

Figure 1 Schematic illustration of LPPD process.

plasma system, an arc of 50-80KW is struck between a W cathode rod and a cylindrical Cu anode shell. Gases (commonly mixtures of Ar or N_2 with He or H_2) flowing around the cathode and through the anode are ionized by the arc discharge, forming a plasma in excess of 10,000°C in temperature. The expansion of the hot plasma leads to supersonic gas velocities at the gun exit.

After the plasma is ignited, the chamber pumping is adjusted to achieve pressures of 50-60 torr near the mandrel. At these pressures the plasma has little interaction with the rarified chamber atmosphere, so that plasma temperatures of 4000-5000°C are maintained even a half meter away from the gun, and plasma gas velocities may be several thousand m/s. The gases have very low density, so that collision with the substrate may heat it to a steady state temperature of 800-1100°C, depending on the substrate size and mass. Such a temperature is desirable to achieve good deposition density for Ni, Co and Fe-base systems. It is possible to include the substrate in the electrical circuit, positive relative to the cathode, to produce electron heating. Reversing the polarity can be used to strip off thin oxide layers present on the substrate to further enhance bonding between substrate and deposit.

Once a near-steady state temperature is achieved at the substrate, powders are injected into the plasma, either in the gun or just outside it (Figure 1). For powders of materials melting in the range of 1300-1600°C, -400 mesh (<37 μm) powders are generally used. These particles are melted and propelled towards the substrate, where they impact and spread prior to being rapidly quenched to the solid state by the cooler substrate or the solidified deposit previously quenched (11,12). From the entrance of the cold particle into the plasma until it is melted, accelerated and deposited, times of 2-3ms are involved (13). At the substrate, spreading occurs in times of the order of 0.1-1μs (the time for the particle to more across a distance equivalent to its diameter of ~25μm at ~250m/s). This leads to cooling rates on the order of 10^6°C/s. Deposition continues until the desired thickness is reached. The plasma is then extinguished, the deposit is allowed to cool, and the chamber pressure is brought to ambient.

Spreading and solidification are shown schematically in Figure 2a for a molten particle. A typical atomized powder screened to -400 mesh (Figure 2b) contains some particles as large as 37 μm, and has an average size of ~25 μm, with ~10% of the particles being smaller than 12 μm. For conditions which melt the coarsest particles, the finest particles may be vaporized. Even some particles not vaporized may be too fine to be separated from the plasma, having insufficient inertia to penetrate the stagnant gas layer above the substrate surface. Temperature and velocity of particles reaching the substrate will depend on particle size and material physical properties, as well as plasma parameters (14-16).

For composite material deposition, the particle size of the second phase is normally chosen to be consistent with plasma parameters governed by the Ni, Co or Fe matrix (3). For a second phase with higher melting point than these materials, finer particles are used. For example, in composites produced with -400 mesh Ni alloys, 15μm particles of alumina are used, since alumina melts at 2050°C; similarly, 9μm particles of Mo are used, since Mo melts at 2617°C. These particles will also spread and solidify at the substrate, but will have smaller dimensions than the splatted discs observed for Ni, Fe and Co particle deposition. Spreading of Ni alloy particles is seen at the outer surface of a deposition in Figure 2c. Also apparent on the surface are numerous small spherical particles. These may be the remnant of a splashing phenomenon as the liquid particles impacted the substrate. Some may also be small particles which condensed and rained from the vapor created by vaporization of extremely fine powders in the -400 mesh screening.

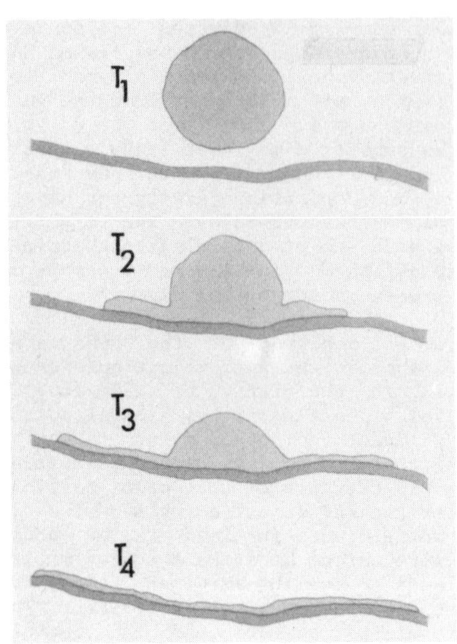

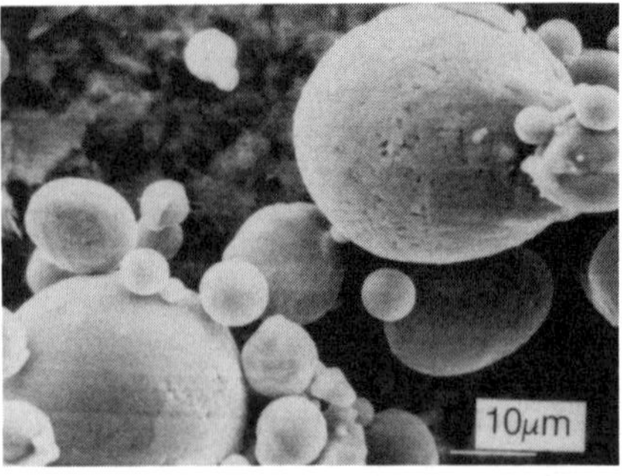

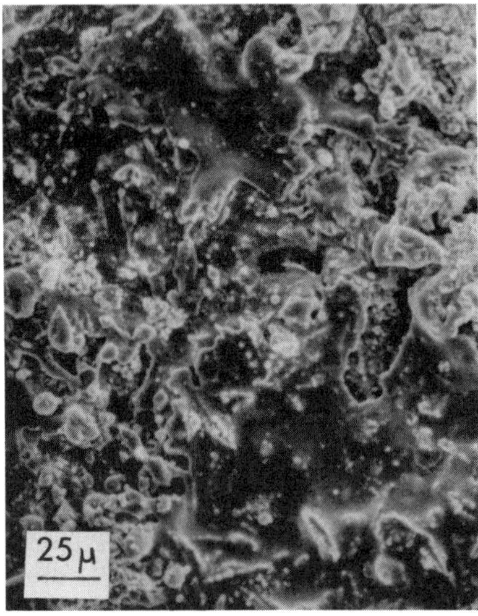

Figure 2 (a) Schematic illustration of liquid droplet spreading as a function of time, (b) comparing -400 mesh Ni alloy powders to (c) the deposited surface.

Structures

For the rapidly cooled Ni, Co or Fe powders, a columnar grain structure results, with grains running across the thickness of the disc-like structure (14). This is seen in Figure 3a. Looking at the plane of the substrate surface, these columnar grains are seen to be equiaxed in that orientation (Figure 3b). The grains are typically 0.2-0.5 μm in cross-section as solidified, with some epitaxy from disc to disc. For matrices of simple composition, there can be considerable grain growth in situ, so that much coarser, non-columnar structures are observed. For the second phase, where smaller size particles are being spread and solidified, many different grain structures have been observed. For example, Mo shows a relatively equiaxed microstructure (17), while TiC shows both columnar and very fine equiaxed grains (16).

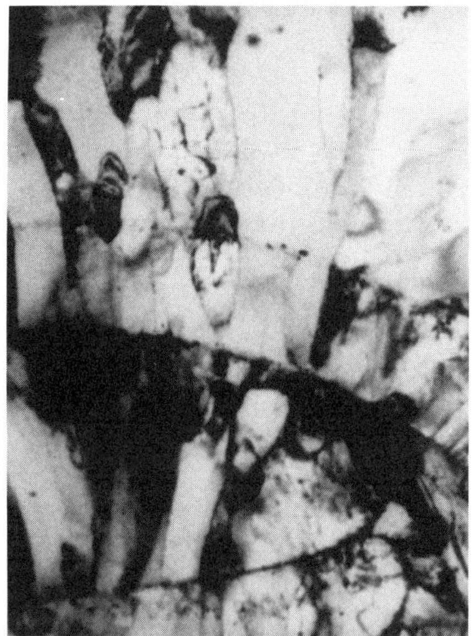

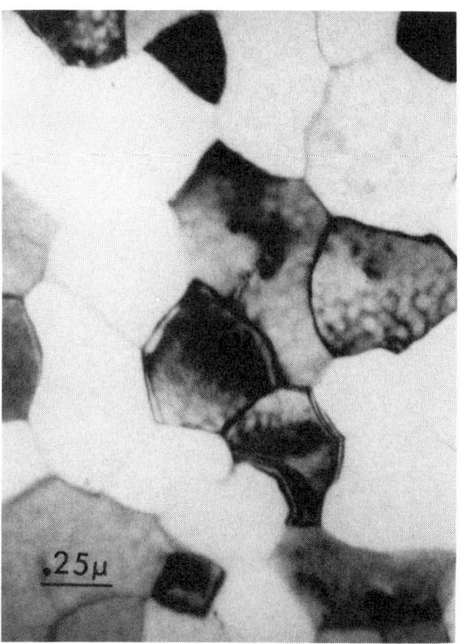

Figure 3 Transmission electron micrographs through (a) plane parallel to spray axis, and (b) plane perpendicular to spray axis.

The densification during heat treatment can be seen for a monolithic superalloy shown in Figure 4. As an as-deposited sample is heated, it expands as would a conventionally produced, fully dense sample of the same alloy. However, at high temperature the microporosity is eliminated due to sintering (11). While sintering is occurring the length of the sample decreases at constant temperature, or at increasing temperature its length increases at a lesser rate than if it were fully dense. The attainment of full density during the high temperature sintering heat treatment is evidenced by the equivalent changes in sample length during the subsequent cooling and reheating cycles. The densification of composite structures occurs in a similar manner. However, during cooling from the sintering temperature, differences in expansion between the two phases may cause a hysteretic cyclic cooling and heating response, due to plastic deformation or cracking of the phase(s).

For monolithic deposits of high temperature materials, the fine grain size is relatively stable during heat treatment, so that grain sizes of 25-250μm are generally the limit achievable (18). Such grain sizes are much smaller than found in typical castings, so that the LPPD materials are generally less resistant to creep than are cast materials (11). Use of the monolithic materials is expected to be as replacements for materials presently used in wrought form: compositions capable of very high strength - which could not be made as wrought materials - can be produced by LPPD in net shape. An example of a net shape monolithic material is shown in Figure 5a. Here a 300 μm thick cylinder 2.5 cm in diameter has been produced, incorporating a wrought flange or embossment in the tubular structure. This was accomplished by attaching the flange to the sacrificial mandrel prior to deposition. After removal of the mandrel, the flange was an integral part of the deposit. Composite structures similar in shape have been made routinely using the LPPD process. An airfoil configuration produced by LPPD processing, shown in Figure 5b, again used sacrificial mandrels to define the deposit shape. Wrought parts were included in the mandrel to form the bridge and trailing edge features.

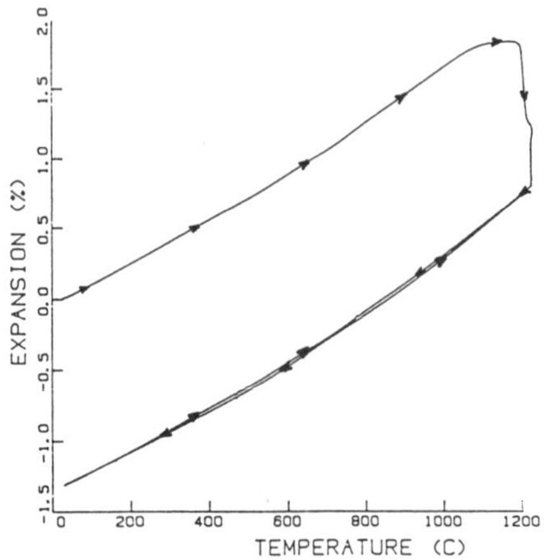

Figure 4 Densification behavior of LPPD structures through heating, cooling and reheating cycle.

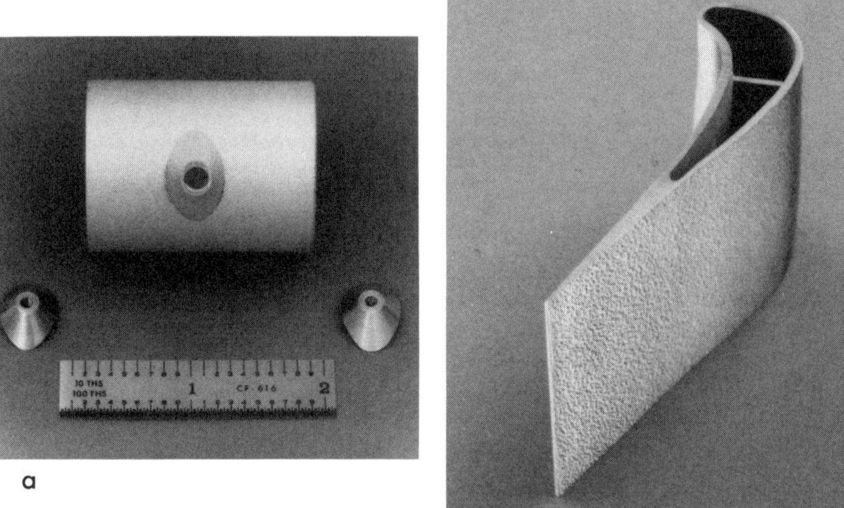

Figure 5 Near net shapes of (a) thin wall tube capturing embossments and (b) airfoil shapes capturing wrought internal structure and trailing edge.

Producing composite structures by LPPD offers an extremely broad range of possibilities for microstructural control (3). Several examples are shown in Figure 6. Structures can be produced which are graded chemically, as in Figure 6a. There a tungsten surface was deposited on the mandrel, and the outer deposit was highly alloyed Ni-base superalloy. Since diffusion of W into the superalloy could cause formation of deleterious embrittling phases, an interlayer was deposited between the W and the superalloy. The layer was composite in nature, consisting of a simple Ni-Al-Mo γ/γ' alloy and α Mo. The α Mo was deposited on the LPPD W layer, then gradually the γ/γ' alloy fraction was increased and the Mo content decreased until only the simple γ/γ' alloy was deposited. Finally, the Ni-base superalloy was deposited over the simple γ/γ' alloy. The W and Mo deposits were harmless to each other as interdiffusion occurred, and the simple and the complex superalloys were also harmless to each other. The Mo and the simple γ/γ' alloy were chemically compatible, and the mechanical compatibility of the entire composite was increased because of the reduction in the rate of change in thermal expansion behavior on traversing the structure from W to the complex alloy.

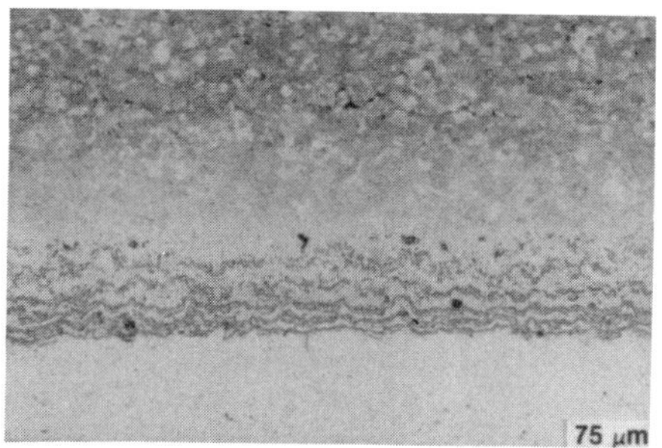

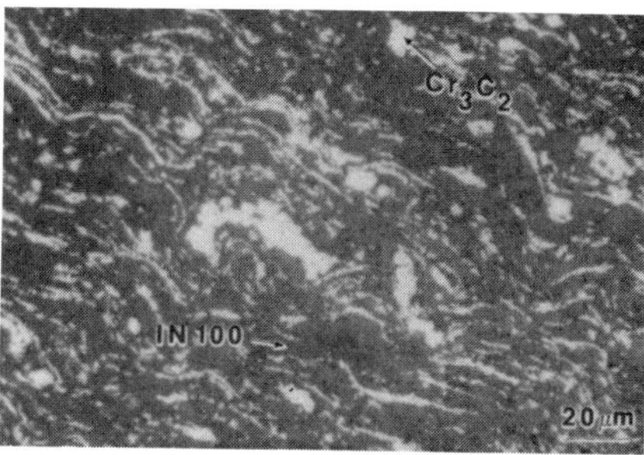

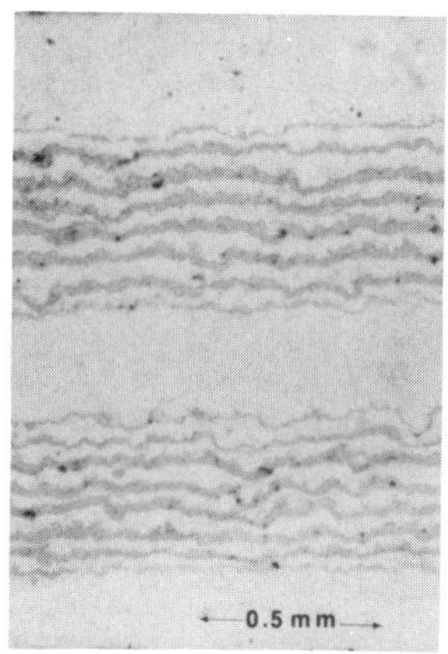

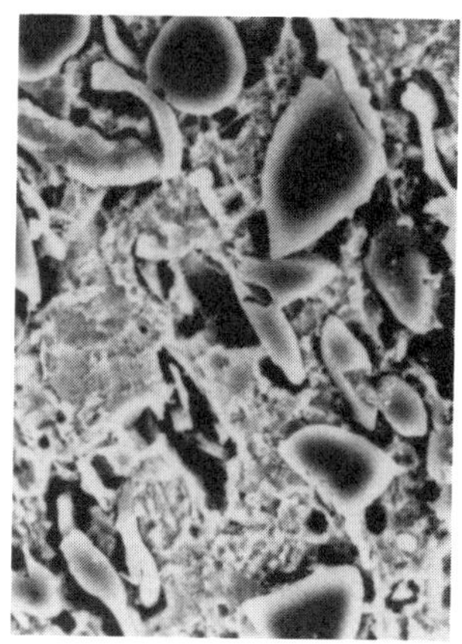

Figure 6 Composite structures made by LPPD

a. grade from W through γ/γ' + αMo to superalloy

b. structure with ductile surfaces and core, with composite sheets of controlled thickness

c. mixed microstructure of TiC and stainless steel 304

d. mixed structure of primarily unmelted alumina particles in a Ni-base alloy matrix.

Another composite, shown in Figure 6b, consists of an Fe-Ni metallic layer and elemental Si (the darker phase). The inner and outer surfaces and the central core consist solely of ductile Fe-Ni metal. The two composite zones have Si layers that are thin adjacent to the fully metallic regions, and Si layers that are thicker in the middle of the composite zones. Both phases are essentially continuous laminates. This structure was produced at relatively low temperatures to avoid reactions involving a liquid phase formation at temperatures >965°C. Variation in laminate thickness could allow control of the positioning of maximum residual stresses in composites.

Two other structures are also seen in Figure 6. A structure where the major phase is TiC and the minor phase is 304 stainless steel, is shown deposited on a steel substrate (Figure 6c), and a Ni-base superalloy with particulate alumina is shown in Figure 6d. The particulate form of alumina was achieved by using a powder size that would not melt under plasma conditions used to melt and deposit the superalloy.

As was noted in the discussion of Figure 6b, the temperature of the deposit can be decreased to avoid low melting reactions. This is accomplished by increasing the distance from gun to substrate to put the substrate in a lower temperature regime of the plasma. However, there is still a question of whether local melting of the deposit can occur. This is shown schematically in Figure 7 (17). For the case of superalloy-Mo composites, the superalloy melting range is more than 1000°C lower than the range for Mo. There is a possibility of Mo liquid particles being quenched on the deposition surface, but still increasing the temperature of the Ni alloy deposited previously to exceed the superalloy-Mo eutectic (>1315°C). This seems most likely to occur when continuous laminations of each phase are being deposited, or when very high volume fraction Mo composites are being produced. Transmission microscopic studies of such composites have produced no evidence that re-melting of the lower melting point material occurs locally. It appears that this event may occur only when the bulk deposit is very close to its melting point, so that any local event may trigger an overall instability with resultant gross melting.

Properties

Both physical and mechanical behavior has been measured for several LPPD composites (3,19,20). Density, elastic, modulus, and thermal expansion behavior, as well as tensile and rupture behavior, will be discussed.

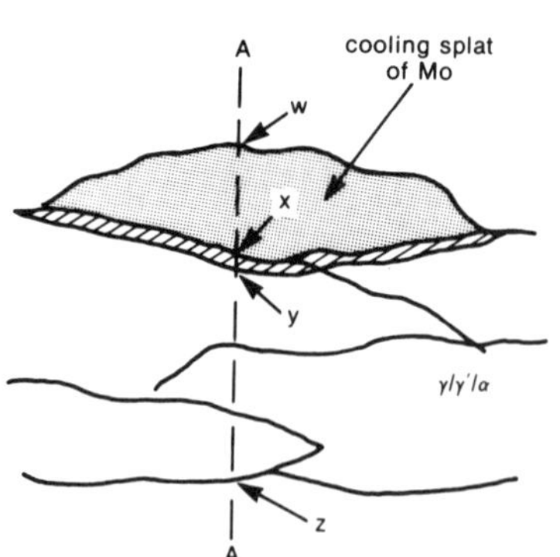

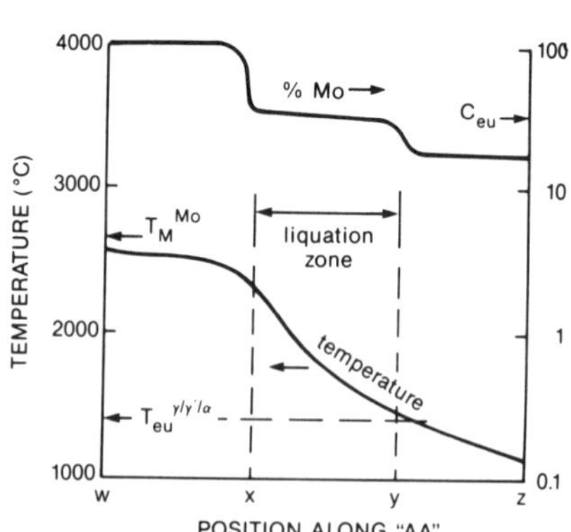

Figure 7 Schematic illustration of temperature and composition for reheating of Ni alloy splats by the arrival of αMo splats possibly leading to partial liquidation of the Ni alloy splats.

Elastic modulus measurements of LPPD composites have been made for several different structures (3,19,20). A simple rule of mixtures description is generally adequate to represent measured data. This is shown in Figure 8 for the Ni-Al-Mo γ/γ'-α composites (20). Modulus is shown both as a function of the nominal Mo volume fraction (Figure 8a) and as a function of the measured density. The better agreement with density - which should vary linearly with volume fraction - indicates that the deposits may contain quite different volume fractions of the phases than is expected from the nominal powder feed rates into the plasma.

This difference between nominal and actual volume fractions can result from one phase being more fully melted than the other, so that more solid particles bounce off the deposit surface. This would produce a systematic difference between nominal and actual volume fractions. Differences in particle size and phase density can also result in injection into different parts of the plasma plume. Resultant changes in volume fraction melted and in particle velocity and particle separation from the plasma at the substrate are also expected to produce systematic variations. Erratic differences between nominal and actual volume fractions may occur because of sudden changes in gun performance, or because of gradual wear of the powder feed parts. Changes in nominal volume fractions achieved by altering powder feed rates can also lead to erratic differences between nominal and actual volume fractions if one or the other phase is particularly difficult to feed (see for example, References 16-20).

In Figure 8 a single point falls far from the rule of mixtures line for modulus vs. nominal volume fraction, but this point is essentially on the rule of mixtures line for modulus vs. measured density (proportional to actual volume fraction). Specific modulus for all the composites is significantly greater than for the matrix alone (ρ = 7.9g/cm^3, E=203 GPa).

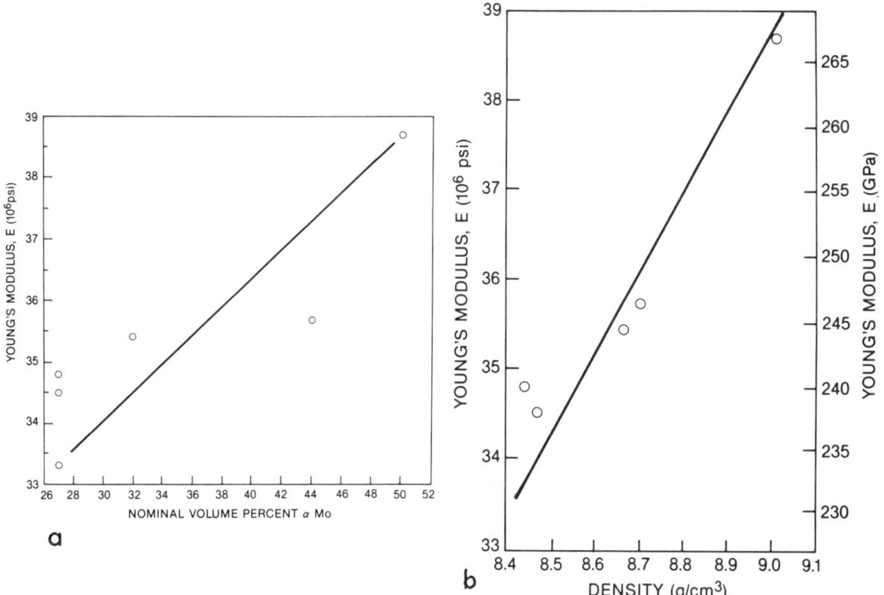

Figure 8 Elastic modulus at room temperature measured in NiAlMo composites as a function of (a) nominal volume percent αMo, and (b) measured density.

Figure 9a compares measured density to nominal volume fraction for the same Ni-Al-Mo composites (20). As would be expected, the same structure shows deviation from rule of mixtures. Only this structure was produced with a high feed rate for a feeder containing solely Mo. An off-nominal particle feed rate may account for the deviation of this point. The measured density of this structure corresponds to a volume fraction of 0.35 Mo, rather than the nominal value of 0.44 Mo.

Expansion behavior for these composites is compared in Figure 9b, for total expansion between room temperature and 1000°C (20). The measured expansions are plotted against the measured densities to avoid any questions regarding the one questionable structure. Variation in expansion is essentially linear with composite density, consistent with a simple rule of mixtures description of expansion. This behavior has been seen for other LPPD composites (3,19). As with any composite system, cycling of the structure through a large temperature range can result in plastic deformation or cracking within the composite. This is evidenced as a hysteretic response to expansion with cycling.

Mechanical properties measurements have been reported for a number of LPPD composite systems (3,19,20). For metal-metal composites standard button-head tensile and rupture bars can be produced readily (20). However, for metal-ceramic materials machining of test bars can present difficulties because high concentrations of refractory oxide can cause rapid machine tool wear. Also, there is the possibility of crack initiation at interfaces between metal and ceramic or within the ceramic due to an excessive materials removal rate. These cracks could be at the surface, or possibly be hidden below the surface.

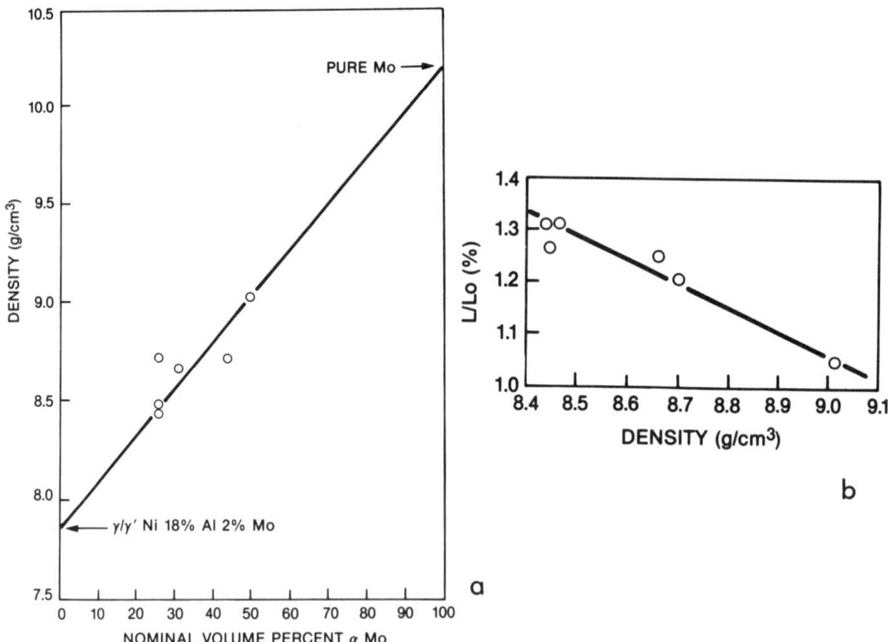

Figure 9 (a) Density vs. nominal volume percent αMo, and (b) thermal expansion at 1000°C vs. density, for Ni-Al-Mo LPPD composites.

A relatively simple and inexpensive test procedure can be used for mechanical testing of LPPD composites (21). A thin ring, 3.8 cm in diameter, 0.1 cm thick, and 0.32 cm wide has been used to test LPPD monolithic and composite materials (Figure 10). Semi-circular cross-section pin-loaded fixtures are placed within the ring, and separated until failure of the ring occurs. The sample is easy to produce, since the deposit can be made to the required thickness on a mandrel of the appropriate outer diameter. Samples can then be sliced or electro-discharge machined to the proper width. This procedure leaves the outer surface of the deposit in the as-deposited condition, with a high degree of surface roughness. This may lead to a slight overestimate in cross-sectional area, and therefore a slight underestimate of composite strength. The rough surface has not led to any obvious tendencies to premature failure. It would be possible to finish the outer surface before slicing, with minimal possibility of damaging the composite.

Mechanical properties of the LPPD composites generally show increased strength and decreased ductility relative to the monolithic matrix. Substantial specific strength increases have been observed. Yield strengths are shown in Figure 11 for FeCrAlY-alumina LPPD composites tested in the ring configuration (19). For structures where alumina and FeCrAlY were deposited as a mixture, yield strength increased monotonically with fraction of alumina, reaching a >three-fold gain at 1010°C for 25 v/o alumina. These structures produced a very significant gain of nearly 100-fold in rupture life.

For the continuous laminate structure (19), which contained approximately 25 v/o alumina (estimated from density measurements and nominal feed rates), no unusual strengthening was observed in tensile tests (Figures 11), but substantially better rupture performance was observed. Continuous laminates in the NiAlMo system showed similar results in rupture (20), but with significantly better performance in tension testing as well.

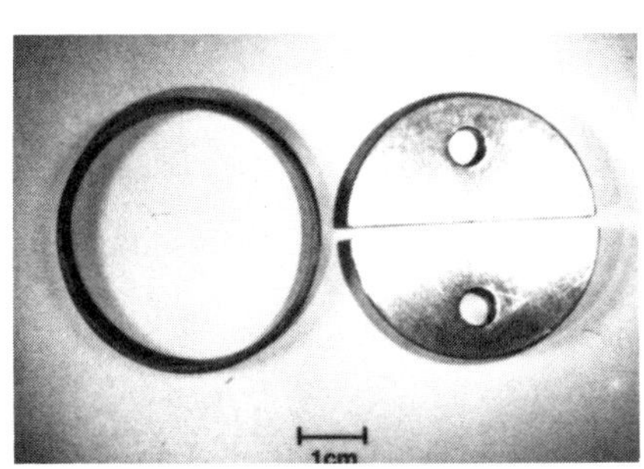

Figure 10

Ring specimen and pin loaded "D" grips for LPPD composite tensile and rupture tests.

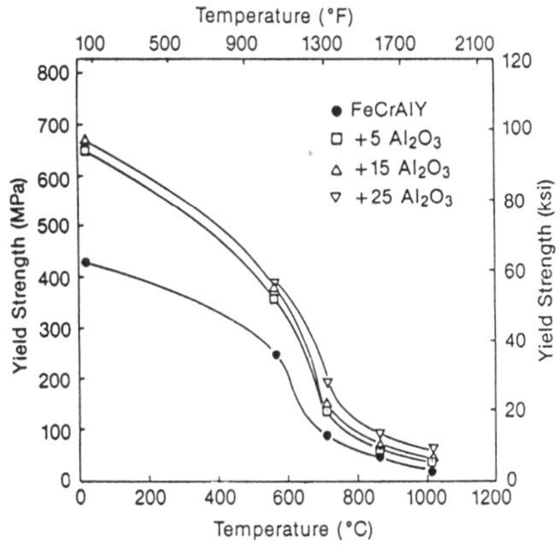

Figure 11

Yield strength as a function of temperature for LPPD FeCrAlY-alumina composites.

Summary

The low pressure plasma deposition process can produce microstructurally sound composite materials capable of significant increases in specific strength and stiffness for high temperature systems. Structures discussed here have been produced with both phases passing through the plasma gun. A rule of mixtures description is an adequate approximation to characterize elastic modulus and thermal expansion behavior.

The process has exceptional flexibility for tailoring the microstructure. Metal-metal, metal-carbide, and metal-oxide structures have been produced over a wide range of volume fractions and distributions of the second phase. Although most of the results available have been generated on systems that might be considered as model materials combinations, the process is capable of producing real engineering materials composites.

References

1. "USAF Promoting Radically Different Design Concepts", Gas Turbine World, Nov.-Dec. 1986, pp. 16-21.

2. M.R. Jackson, J.R. Rairden, J.S. Smith, and R.W. Smith, and R.W. Smith, J. Metals 33, 1981, pp. 23-27.

3. P.A. Siemers, M.R. Jackson, R.L. Mehan, and J.R. Rairden, III, Ceram. Eng. and Sci. Proc. 6, 1985, pp. 896-907.

4. G.M. Giannini, Sci. Amer., August 1967, p. 2.

5. T.A. Roseberry and F.W. Boulger, "A Plasma Flame Spray Handbook", Rpt. No. MT-043, Naval Sea Systems Command, Dept. of the Navy, March 1977.

6. D.R. Mash and I. MacP. Brown, Matls. Engr. Quart., February 1964, p. 18.

7. R.W. Cahn, Proc. Int. Conf. on Rapid Solidification Processing, Reston, Virginia, 1977, p. 129.

8. E. Muehlberger, Proc. 7th Int. Thermal Spray Conf., London, 1973, p. 245.

9. J.R. Rairden, M.R. Jackson, M.F.X. Gigliotti, M.F. Henry, J.R. Ross, W.A. Seaman, D.A. Woodford, and S.W. Yang, Proc. Int. Conf. on High Temperature Corrosion, NACE, San Diego, 1981, p. 72.

10. A.M. Johnson, J.S. Kelm, R.W. Smashey, D.V. Rigney and T.G. Wakeman, in Proc. Third Conf. on Rapid Solidification Processing, NBS, Gaithersburg, MD, 1982, pp. 650-661.

11. M.R. Jackson, R.W. Smashey and L.G. Peterson, in Proc. Third Conf. on Rapid Solidification on Processing, NBS, Gaithersburg, MD, 1982, pp. 198-208.

12. T.L. Cheeks, M.E. Glicksman, M.R. Jackson and E.L. Hall, in Proc. Third Conf. on Rapid Solidification Processing, NBS, Gaithersburg, MD, 1982, pp. 118-123.

13. G. Frind, C.P. Goody, and L.E. Prescott, Int. Sym. Plasma Chem.-6 Montreal 1983, pp. 120-125.

14. A.M. Ritter and M.R. Jackson, in Proc. Third Conf. on Rapid Solidification Processing, NBS, Gaithersburg, MD, 1982, pp. 270-275.

15. D. Apelian, private communication, 1986.

16. A.M. Ritter, M.R. Jackson, and R. Wright - this volume.

17. M.R. Jackson and A.M. Ritter, 86CRD250, General Electric, Schenectady, NY, 1987.

18. M.R. Jackson and R.L. Mehan, to be published in Proceedings, Sixth International Conference on Composite Materials, London, 1987.

19. M.R. Jackson and P.A. Siemers, to be published in Proceedings, Plasma Technology Symposium, MRS, 1987.

20. R.L. Mehan, M.R. Jackson and J.R. Rairden, to be published in J. of Materials Science.

PLASMA-SPRAYED STAINLESS STEEL-CARBIDE PARTICULATE COMPOSITES

Ann M. Ritter, Melvin R. Jackson and Roger N. Wright*

General Electric Company
Corporate Research and Development
Schenectady, New York 12301

*Department of Materials Engineering
Rensselaer Polytechnic Institute
Troy, New York 12181

Abstract

The microstructure and properties of plasma-sprayed 316 stainless steel/carbide particulate composites were studied over a range of 0-40 vol/o TiC and NbC. The matrix grain size varied with carbide content, and the grain size within carbide lamellae was a function of carbide type. The degree of carbide melting during spraying was related to the total amounts of carbide in the powder mixture. The yield stress increased with carbide additions, and the magnitude of the increase was greater for TiC than for NbC additions. Increases in the yield stress were related to decreases in the mean free path between the particles. The ductility decreased with carbide additions, and the extent of the ductility loss was related to carbide size distribution, carbide type, and the rate of carbide cracking. The presence of very large particles resulted in low tensile ductilities at room temperature. The ductilities of NbC-containing samples were generally higher than for TiC-containing specimens. Fracture of the composite occurred when specific amounts of cracked carbide were present. Therefore, the macroscopic fracture strain was increased by delays in carbide cracking to higher applied strains. Delay in carbide cracking was observed in room temperature tensile samples heat-treated and quenched instead of HIP-consolidated, and in 760°C tensile specimens.

Processing and Properties for Powder Metallurgy Composites
Edited by P. Kumar, K. Vedula and A. Ritter
The Metallurgical Society, 1988

Introduction

Plasma-spraying is a rapid solidification process which has been used to produce coatings[1,2] and structural parts[3] of both metals[4-6] and ceramics[7,8]. Early work[9] was done in air or inert gas, and the resulting deposits were generally high in oxygen and of low density (75-95% of theoretical density, depending on the conditions). With the advent of the low-pressure plasma-spray process[1], as-deposited densities of 96-99% of theoretical values were achieved, and oxygen contents were low. As with other rapid solidification techniques, low pressure plasma deposition can result in structural and microchemical homogeneity[3]. It can be used to produce fine-grained deposits from standard alloys using prealloyed powders. In addition, powders of dissimilar chemistries, such as metals and oxides or carbides, can be sprayed to produce composite structures[1,10,11], in which a ductile matrix contains a random or layered distribution of a hard second phase. Composite coatings fabricated by plasma-spraying have been used extensively for wear-resistant applications[10,12]. Plasma-sprayed composite structures have been produced using the same process[11,13]. The flexibility of the technique is such that both particulate and lamellar composites may be produced by varying the process parameters.

Microstructural characterization and some mechanical property evaluation of plasma-sprayed composites have been reported[11,14]. However, data are not available on the variation of properties with systematic increases in the volume fraction of particulate. Such studies have been done on particulate composites produced by conventional powder metallurgy techniques[15,16], and have related properties to models used to predict strength[17-19]. Comparison of the results with properties of the plasma-sprayed composites is of interest.

The increased strengths attributed to large amounts of second phase particles in composites are accompanied by a rapid loss in ductility[15]. Observations in some alloys[15] indicated that the ductility loss is not affected by the type or size of the particles but only by the volume fractions. However, other investigators[20-22] have shown that both particle size and shape may affect the degree of the ductility loss. To date, this has not been fully characterized.

In the present study, plasma-sprayed particulate composites with TiC or NbC as the dispersed phase are examined. The ductile matrix selected was 316 stainless steel, a well-characterized alloy with wide use. The effects of carbide additions on strength are evaluated, and the data related to available models. In addition, the carbide cracking behavior as a function of strain, carbide size and test temperature is quantified, and the results related to variations in the ductility.

Experimental Procedure

The materials used were produced by plasma-spraying mixtures of 316 stainless steel powder and carbide powder. Two types of stainless steel powder were used, one -325 mesh water-atomized lot and one -400 mesh gas-atomized lot. The TiC powder lots were one screened to -325 mesh, one with a nominal average particle size of 3μm, and one with a nominal average particle size of 1μm. These will be referred to as -325 TiC, 3μm TiC and 1μm TiC. One NbC powder lot with a nominal average particle size of 2μm was also used.

The steel and carbide powders were mixed together and plasma-sprayed with one gun. The spraying was done in a low-pressure plasma system described in this volume[13]. The various parameters are listed in Table 1. The powders were sprayed onto a cold-rolled steel plate, which was dissolved after deposition. The deposits were generally ~98% of theoretical density, and further densification was achieved by either heat-treating in vacuum for 4 hours at 1200°C followed by a water quench, or hot- isostatically-pressing (HIP) for 1-2 hours at 1100°C and 170MPa. All 1μm TiC and NbC samples were tested in the HIP'ed condition.

TABLE 1

Plasma-Spraying Parameters

Plasma Gas Parameters	Ar: 1.4 l/s He: 0.3 l/s
Gun Voltage	42V
Gun Amperage	1900A
Chamber Pressure	8k Pa (60 torr)
Powder-Feed Rate	1 g/s
Powder Injection	Internal; 20° angle
Carrier Gas	Ar: 0.12 l/s
Gun to Substrate Distance	30 cm (12")
Gun Traverse Rate	15 cm/s (6"/s)

Microstructural analysis was done by optical metallography on polished samples, and by transmission electron microscopy on thin foils. The foils were electrochemically polished using a solution of 20% perchloric acid/80% methanol at a temperature of about -10°C, with polishing conditions of 25V and 80-100mA. The thin foils were examined with a scanning-transmission electron microscope equipped with an energy dispersive analyzer.

Tensile tests were performed using samples 0.25cm (0.1") diameter in the uniform gauge section, and with effective gauge lengths of 1.42cm (0.56") or 1.08cm (0.425"). All specimens were machined such that the loading axis was perpendicular to the direction of plasma deposition. Tests were done at room temperature in air and at 600°C and 760°C in a vacuum of about 0.01Pa ($1x10^{-5}$ torr).

Monotonic bend tests were run using metallographically-polished samples loaded in three-point bending. Replicas of the polished surfaces were taken at each strain increment. The amount of cracked carbide at each strain was determined from micrographs taken on the replicas.

The size distribution of carbides in the deposits was determined by dissolving the matrix and examining the extract in a scanning electron microscope. The major axis of each particle was measured, and an average value calculated. This was the value of average particle size, d_p, used in the equation[15] for determining the mean particle spacing, D_p, in a sample as follows:

$$D_p = d_p \ (1-f) \sqrt{\frac{2}{3f}} \qquad (1)$$

where f is the carbide volume fraction. Similarly, the mean free path, λ, between carbides in a sample was calculated from

$$\lambda = \frac{2d_p}{3f} \ (1-f) \qquad (2)$$

The mean free path is the distance from a reference particle to a second particle, averaged in all directions[15]. Interparticle spacing is the average diameter of space-filling spheres, with dispersed particles as centers[15]. The formulas used assume spherical particle shapes.

Results and Discussion

1. Composite Microstructure

The average particle size and particle size distributions for the -325 and 3μm TiC powder extracted from deposits are shown in Figure 1. The particle size distribution for both lots was broad, ranging from less than 0.5μm to a maximum of 35-50μm. The main difference between the two lots was the presence of more of the large (>10 μm) particles in the -325 TiC. The narrower size distributions for the 1μm TiC and NbC are shown in Figure 2. There were very few large (>10μm) particles in the NbC, and no particles larger than 5μm in the 1μm TiC.

The morphology of the TiC and NbC powder particles can be seen in Figure 3, which shows deposits containing 7-9 vol/o carbide. Very little melting of the carbide in the plasma occurred. The angular shape of the TiC particles can be contrasted with the somewhat more rounded NbC particles. Both of these deposits were made from powder mixtures containing 10 vol/o carbide. When $\geq$20 vol/o carbide was in the mixture, nearly all of the TiC tended to melt (Figure 4), resulting in a lamellar microstructure. A few large unmelted particles were always observed. When the powder mixture contained $\geq$40 vol/o carbide, extensive melting of the NbC particles occurred.

Transmission electron microscopy showed very fine grains within the TiC lamellae (Fig. 5), with both equiaxed and columnar grains. The TiC grain size in these lamellae was on the order of 20-100nm. By comparison, the grains in the NbC lamellae were about 0.5μm in size, and were generally equiaxed (Fig. 6). This dependence of grain size on carbide type was also observed in specimens made from powder mixtures containing both TiC and NbC.

The grain size in the 316 stainless steel matrix varied with the amount of carbide present in the deposit. In specimens containing small amounts (<10 vol/o) of carbide, the average matrix grain size was on the order of 10μm. In samples containing 20-40 vol/o carbide, the average matrix grain size was about 1-2μm. The variability in matrix grain size can be attributed to inhibition of grain growth by second-phase particles. During plasma deposition, the solidified material was at a high temperature (900-1000°C) while the plasma plume traversed the deposit. After solidification, matrix grain growth occurs in single-phase materials[23]. Materials in which second-phase particles form during solidification tend not to show appreciable matrix grain growth, since this is impeded by the presence of precipitates on cell or grain boundaries. As a result the solidification structure in plasma-sprayed samples of such materials as Ni-base superalloys is generally preserved, with grains commonly 0.25-0.5μm in size[3,14,23]. The carbide particles in the 316SS/carbide composites also limited grain growth in the stainless steel, but not to the same extent.

The grain size variation between different types of carbide lamellae may be the result of different cooling rates of the lamellae subsequent to solidification. It has been shown[24,25] that the cooling rate increases as the lamella thickness and density decrease. The NbC lamellae tended to be thicker than the TiC lamellae, and the density of NbC is significantly higher than TiC. This may have resulted in a slower cooling rate for the NbC lamellae after solidification, allowing growth of the NbC grains.

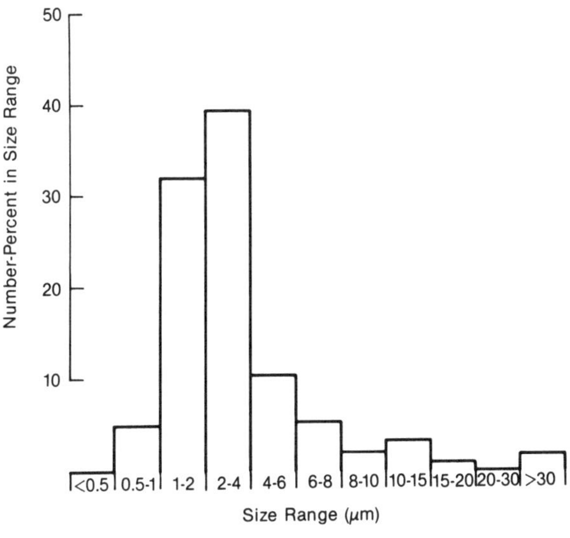

a

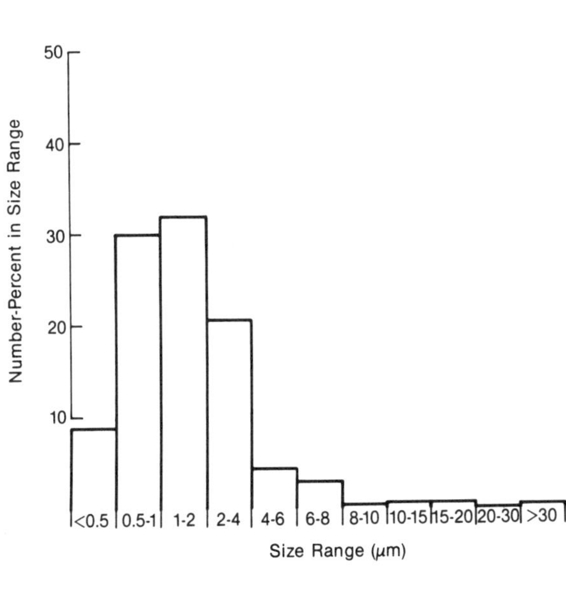

b

Figure 1. Particle size distribution in extracted carbides from deposits containing (a) -325 TiC, and (b) 3μm TiC.

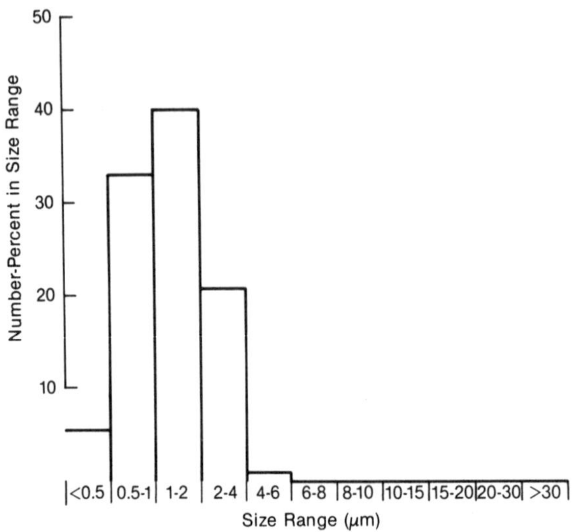

a

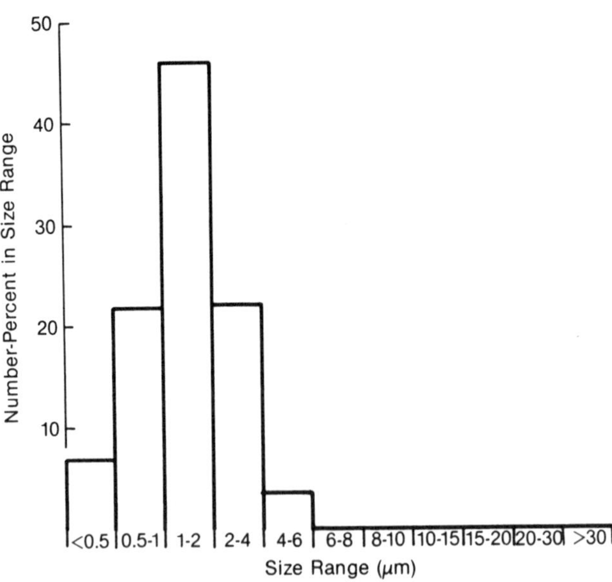

b

Figure 2. Particle size distribution in extracted carbides from deposits containing (a) 1μm TiC, and (b) NbC.

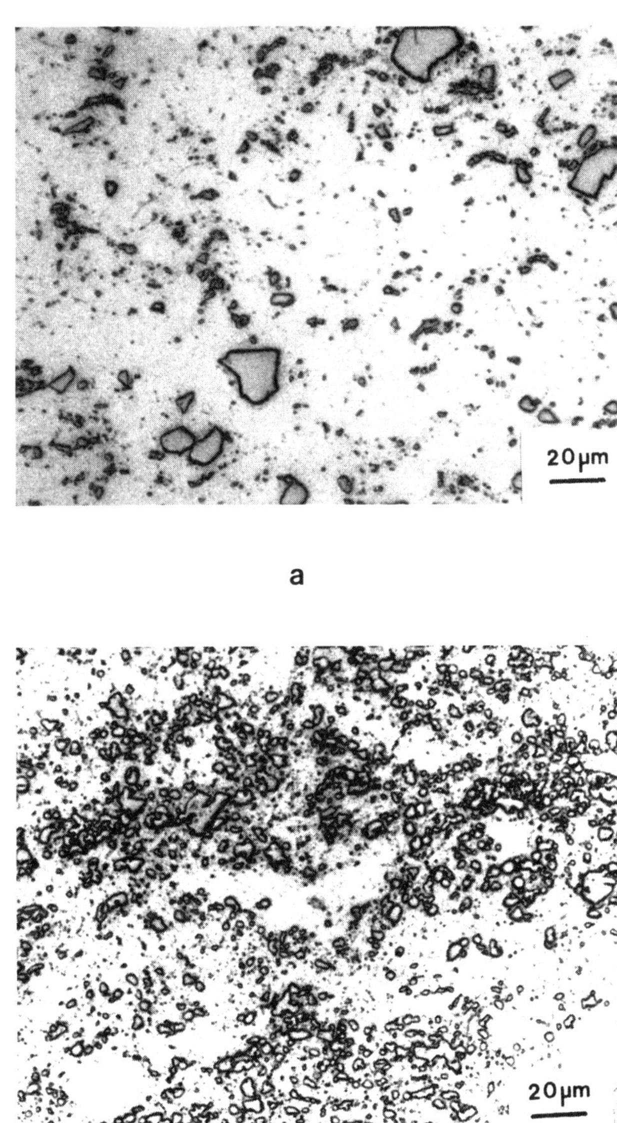

Figure 3. (a) Angular, unmelted TiC particles in deposit containing 7 vol/o TiC; and (b) NbC particles in deposit containing 9 vol/o NbC.

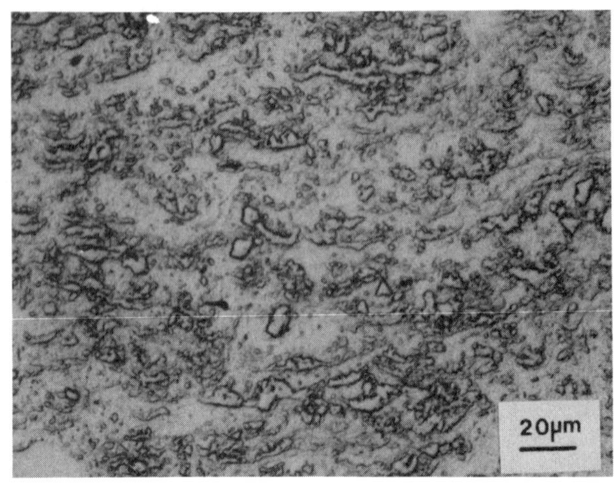

Figure 4. Melted, lamellar TiC in deposit containing 22 vol/o Tic.

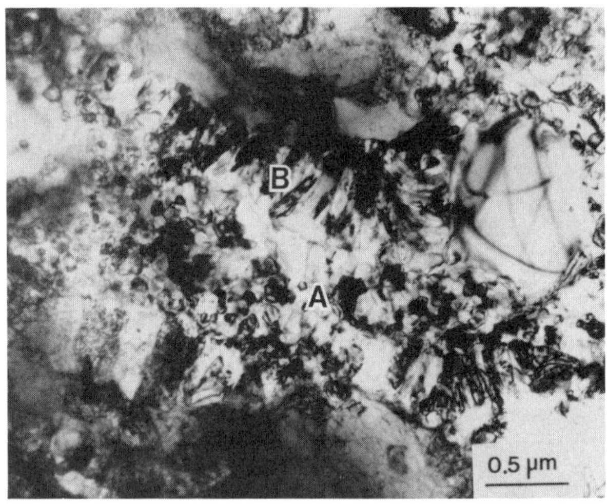

Figure 5. Equiaxed (A) and columnar (B) grains in TiC lamella, in as-HIP sample containing 28 vol/o TiC.

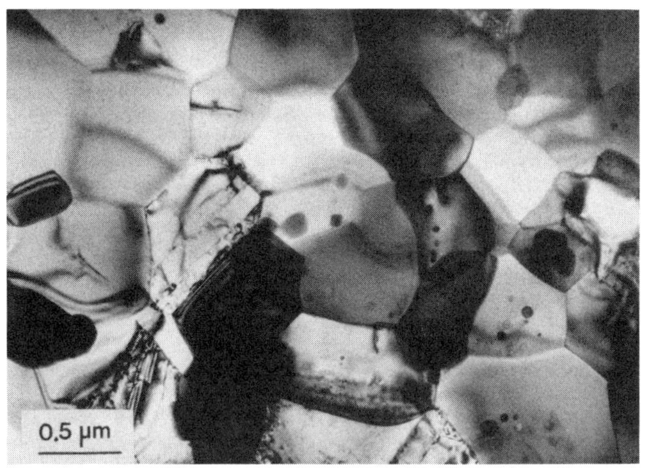

Figure 6. Equiaxed grains in NbC lamella in as-HIP sample containing 40 vol/o NbC.

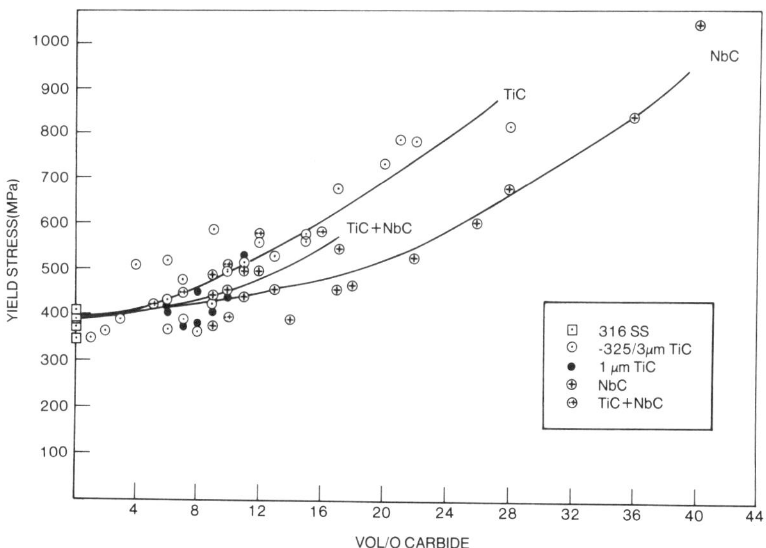

Figure 7. Effect of carbide additions on room temperature yield stress in samples containing TiC, NbC or mixtures of TiC and NbC.

2. Composite Properties

Yield Stress

The general effects of carbide additions on the composite properties were to increase to flow stress and work-hardening rate, and to decrease the fracture strain. The effect of increasing amounts of carbide on room temperature yield stress is shown in Figure 7. The curves drawn through the data are for samples containing only TiC, only NbC or mixtures of TiC and NbC. The strengthening effects of -325 TiC, 3μm TiC and 1μm TiC were comparable at the same volume fraction of carbide. Specimens containing TiC were generally stronger than those containing NbC. For all samples containing small (<10 vol/o) amounts of carbide, some strengthening was observed. Large increases in strength resulted when large amounts of carbide were present. Similar trends were seen in samples tested at 600°C and 760°C.

The presence of rigid inclusions in a work-hardening matrix has been predicted[18] to increase the strength, as long as the particle/matrix bond is strong[15] and this effect has been reported for several alloy systems[15,19,26]. The yield strength of particulate-containing materials has been commonly related[19,23,27,28] to the logarithm of the reciprocal mean free path, $1/\lambda$, and to the reciprocal of the interparticle spacing (Eqn. 1,2). The data for the 316SS/carbide composites exhibited the general trends observed in other particulate materials. As seen in Figure 8, there was little strengthening below a critical value of $1/\lambda$, and linear hardening over a range of $1/\lambda$ values at room temperature. These trends were also observed at 760°C. For a specific volume fraction of carbide, the NbC and 1μm TiC samples had a higher $1/\lambda$ value than the -325 and 3μm TiC samples. If the strength were a function of reciprocal mean free path only, one would expect the 1μm TiC and NbC samples to be stronger than the -325 and 3μm TiC specimens. However, the tensile data (Fig. 7) indicated that all TiC samples had approximately equivalent strengths, and that the NbC samples were somewhat weaker than the TiC specimens. This discrepancy between the model and material behavior was also seen when the yield stress was plotted as a function of the Orowan-Ashby parameter, which includes the particle size and the reciprocal of interparticle spacing[27].

Edelson and Baldwin[15] found a correlation between increasing yield strength and increasing volume fraction of particulate in composite materials. However, the data for different types of particulate tended to lie on different curves. When the yield stress values were plotted against the logarithm of the reciprocal mean free path, the data fell on one curve for a large range of λ. They concluded[15], therefore, that yield strength was primarily a function of reciprocal mean free path. In the present study, the divergence of the data from this behavior may be due to inaccurate calculation of mean free path and interparticle spacing. The formulas used assume a spherical particle shape and a random particle distribution. Non-spherical morphologies would result in changes in the mean free path[29], and should influence the strength. Kelly[30] has shown that rod-shaped and aligned plate-like particles result in more strengthening than spheres for identical volume fractions of particles. The effect of an elongated shape was to reduce the local mean free path in a slip plane, making bypassing of the particles by dislocations more difficult. In the 316SS/carbide composites, the TiC particles were more angular and elongated than the NbC particles. In addition, many of the TiC specimens contained lamellae, while the NbC specimens generally did not. The disc-like shape of the lamellae would be expected to influence the mean free path, and the combination of this and particle angularity may be the primary explanation for the higher strengths of the TiC samples relative to the NbC specimens.

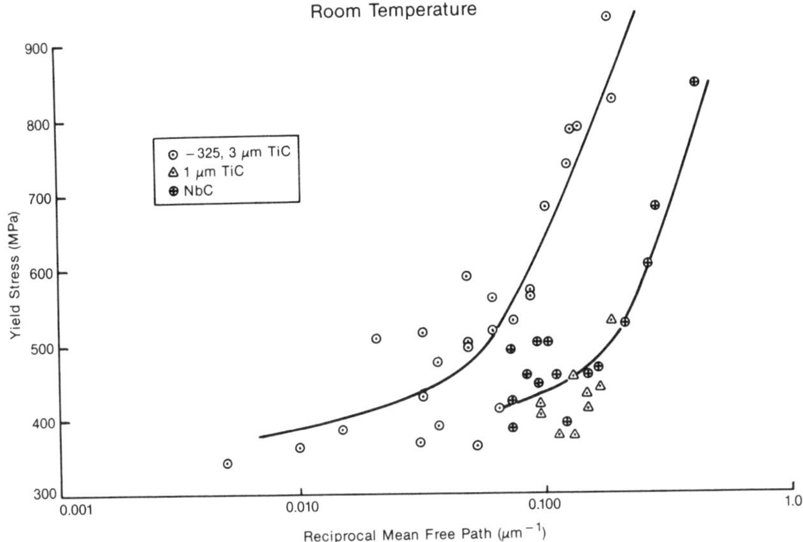

Figure 8. Variation of room temperature yield stress with reciprocal mean free path in TiC and NbC tensile samples.

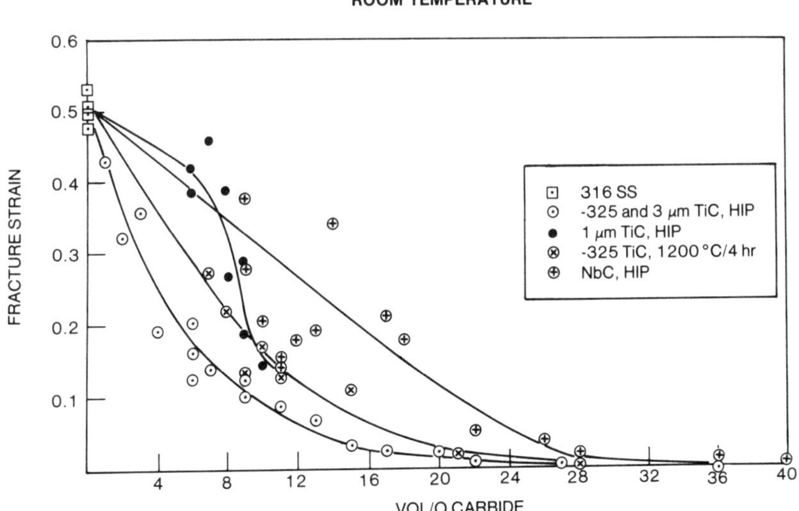

Figure 9. Effect of carbide additions on room temperature fracture strain in samples containing TiC or NbC.

Additional error in calculating the mean free path may have arisen from the non-random, layered distribution of the particles in these composites. Carbide stratification was very pronounced in the 1μm TiC samples. This may account in part for the discrepancy between the predicted higher strength of these samples, and the observation that they had strengths comparable to the -325 and 3μm specimens.

Ductility

The effect of increasing amounts of carbide on the true fracture strain at room temperature is shown in Figure 9. The rapid decrease in fracture strain with small additions of TiC is consistent with observations made on other particulate systems[15,31], and the scatter in the data is typical for particulate composites. The four curves drawn through the data show distinct differences between sample types. The -325 and 3μm TiC samples which were HIPed were the least ductile, while -325 and 3μm TiC specimens which had been heat-treated and water-quenched were more ductile than the HIP'ed samples over the range of 7-15 vol/o carbide. The 1μm TiC HIP'ed specimens were significantly more ductile than the -325 and 3μm HIP'ed TiC samples for the range of structures tested (6-11 vol/o), and were more ductile than the heat-treated -325 and 3μm TiC samples over most of that range. The NbC samples were generally more ductile than all TiC samples. At 760°C, the addition of carbide resulted in decreased ductility (Fig. 10), but to a lesser degree than at room temperature. The differences in ductility will be discussed after the cracking behavior of the carbides has been examined.

The ductility loss due to the presence of rigid particles has been attributed[15] to strain concentrations in the vicinity of the particles, which result in the matrix reaching the fracture strain in a local region for smaller macroscopic strains. The presence of cracks in second-phase particles would result in further strain concentration. Particle cracking has been frequently observed[20,32,33] in particulate composites, and cracked carbides were present on the fracture surfaces of the 316SS/carbide material (Fig. 11). Many of the carbides contained multiple cracks, and observations made on bend bars showed that the cracking occurred sequentially with increasing strain. The data from several bend bars are shown in Figure 12, and the cracking behavior was found[34] to be identical for tensile samples and bend bars at carbide levels of ≤15vol/o. The amount of cracked carbide increased with increasing strain. In addition, at any strain level, the volume-percent of cracked carbide was greater in samples containing large total amounts of carbide. Cracking occurred first in the largest carbides, and cracking of smaller carbides was observed as the applied strain increased. This is consistent with results reported[20,26,33] for other systems. Composite fracture occurred in the room temperature -325 and 3μm TiC tensile samples when 5.0±1.1 vol/o carbide was cracked, and when 6.7±1.9 vol/o carbide was cracked in the NbC tensile samples. In the 760°C tensile samples, fracture occurred at 2.2±0.7 vol/o and 2.0±0.2 vol/o cracked carbide for the -325 and 3μm TiC, and NbC respectively. In all samples, there was no observable lateral void growth in the matrix around the cracked carbides, and no decohesion of the particle/matrix interface. Fracture occurred when strain concentrations at the tips of cracks in the carbides initiated unstable ductile tearing in the matrix regions.

Fracture in these composites was related to the particle size distribution, and to the rate of carbide cracking with applied strain. These effects will be discussed below. Local strain concentrations in the vicinity of cracked carbides would be a function of the crack length, and, therefore, of carbide size. The generally lower ductilities of the -325 and 3μm TiC samples relative to the 1μm TiC and NbC samples resulted from the presence of very

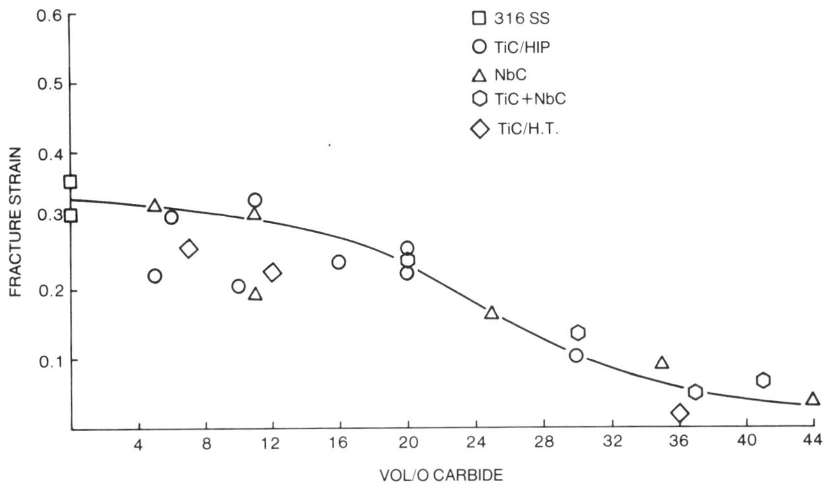

Figure 10. Effect of carbide additions on 760°C fracture strain in samples containing TiC, NbC or mixtures of TiC and NbC.

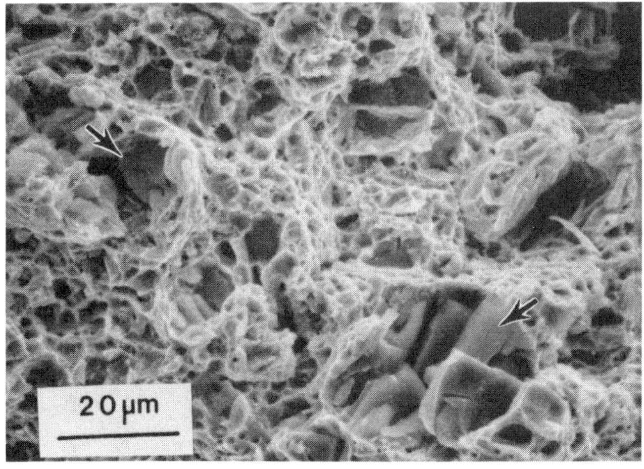

Figure 11. Fracture surface of room temperature tensile sample containing 7 vol/o TiC.

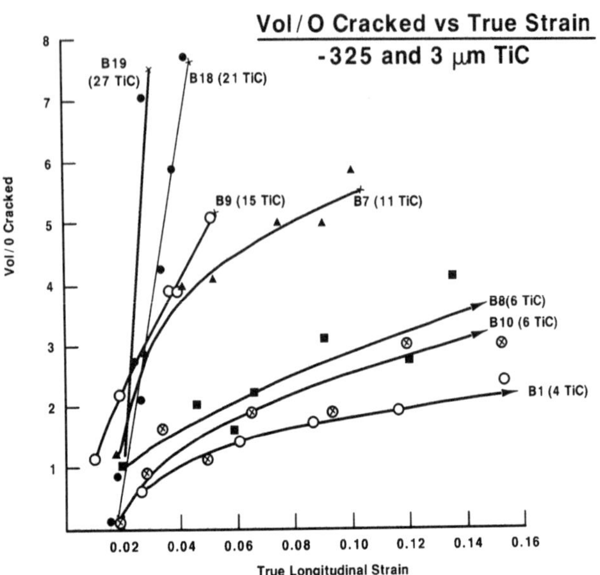

Figure 12. Variation of vol/o cracked carbide with applied strain for bend bars containing -325/3 m TiC. Samples B7, B9, B18 and B19 fractured during the test. Samples B1, B8 and B10 did not fracture, and the curves extend to higher strains.

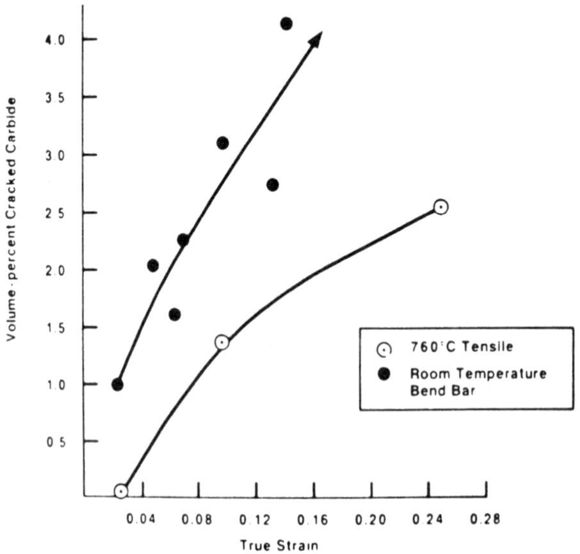

Figure 13. Variation of vol/o cracked carbide with applied strain for a room temperature bend bar containing 6 vol/o TiC , and three 760°C tensile samples containing 6-7 vol/o TiC.

large carbides (30-50μm) in the -325 and 3μm TiC specimens. These were found to crack at very low applied strains (~0.7%). High strain concentrations associated with these large cracks promoted composite fracture at low macroscopic strains. Evidence of a similar particle size-effect on ductility has been shown in steels[22]. Although the large particles in the -325 and 3μm TiC samples were a small percentage of the total number of carbides (Fig. 1), they had a dominant effect on the ductility.

Particle shape may also affect ductility, as has been seen in steels[21] and Al-alloys[31]. The lower ductility of the 1μm TiC samples relative to the NbC samples may have resulted from strain concentrations associated with the angular morphology of the TiC particles. Alternatively, the NbC particles may be under a different residual stress state than the TiC particles, since the thermal expansion coefficients are somewhat different[35].

Since fracture in the 316SS/carbide composites occurred at critical levels of cracked carbide, delay of carbide cracking to higher strains would enhance the ductility. This effect was seen in the comparison of HIP'ed with heat-treated and quenched -325 and 3μm TiC samples, with the latter more ductile. The data for carbide cracking in samples containing 5-6 vol/o TiC are shown in Table 2, which gives the applied strains at which cracks in carbides were first observed. It is apparent that, in the sample which was heat-treated and quenched after HIP'ing, cracking occurred at higher strains for all carbide size ranges examined. This delay in carbide cracking was probably the result of higher compressive residual stresses in carbides in the heat-treated sample due to the faster cooling rate. The thermal expansion mismatch between TiC and stainless steel would result[36] in compressive residual stresses in the carbide after cooling to room temperature, and the magnitude of the compressive stresses would increase with increasing cooling rate from elevated temperatures. Compressive residual stresses in the carbides would oppose tensile stresses during testing, and delay carbide cracking to higher applied stresses and strains.

TABLE 2

Particle Cracking: 5-6 TiC

Carbide Size	Strain for First Crack	
	HIP	HIP+H.T.
40-50μm	0.007	0.013
10-20μm	0.011	0.023
5-10μm	0.018	0.032

Delayed carbide cracking was also one factor in the enhanced ductility of the composites at 760°C, relative to room temperature. The delayed cracking is seen in Figure 13, which shows data from a room temperature bend bar and three 760°C tensile samples. Stresses in the carbide which cause cracking are a function of the applied strain and of matrix work-hardening[18]. The low work-hardening capability[34] of the stainless steel at 760°C would result in lower tensile stresses in the carbide at a specific level of applied strain, and this would delay carbide cracking.

Some of the enhanced ductility at 760°C may be the result of a small amount of carbide plasticity, and of a change in the matrix deformation mode. Examination of carbides in HIP'ed material and in room temperature and 600°C tensile specimens showed little dislocation substructure. In 760°C tensile samples, many of the carbides contained dislocations, an example of which is

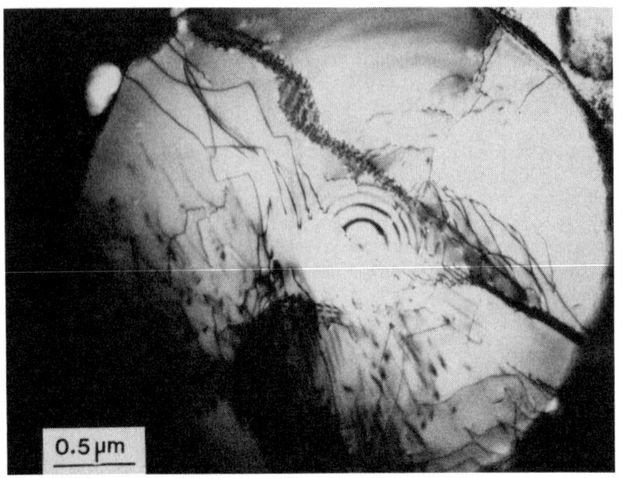

a

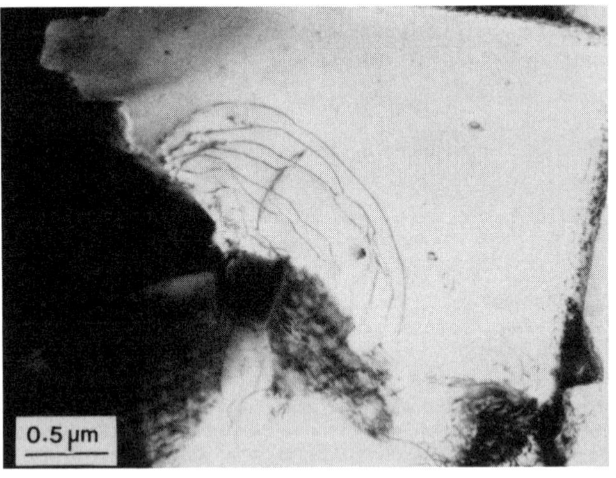

b

Figure 14. Dislocations in carbide particles in 760°C tensile samples containing (a) 11 vol/o NbC, and (b) 11 vol/o TiC.

shown in Figure 14. The onset of a small amount of plasticity in the carbide could increase its fracture resistance. In addition, deformation of the matrix at room temperature occurred both by dislocation movement and by twinning while twins were generally absent at 760°C. The interaction of twins with carbides at room temperature may have contributed to carbide cracking, as has been seen in carbide fibers[37].

CONCLUSIONS

(1) Plasma-sprayed 316 stainless steel/carbide composites consist of a fine-grained steel matrix with both melted and unmelted carbides. The degree of carbide melting is a function of the amount of carbide in the powder mixture. Grain size within the melted carbide lamellae is a function of carbide type.

(2) The addition of up to 40 vol/o carbide to 316 stainless steel increases the yield strength at room temperature, 600°C and 760°C. These increases are approximately related to the decreasing mean free path between the particles.

(3) The addition of TiC has a more pronounced effect on strength than NbC. This may be related to local mean free path decreases arising from the angular morphology of the TiC particles.

(4) Composite ductility is decreased by increasing carbide additions. The fracture strain is also a function of particle size distribution, with the presence of very large particles having a deleterious effect on ductility.

(5) The type of carbide has an effect on ductility, with NbC-containing specimens generally more ductile than TiC-containing samples. This may be a function of higher strain concentrations associated with the angular TiC particles, or of different residual stresses in the two types of carbide.

(6) One of the requirements for composite fracture was the presence of a critical amount of cracked carbide. Delay of carbide cracking to higher strains results in larger composite ductilities. Carbide cracking can be delayed by the presence of higher residual compressive stresses in the carbides, or by low matrix work-hardening rates, as seen at 760°C.

References

1. E. Muehlberger, Proc. 7th Int. Thermal Spraying Conf., British Welding Society, London, England (1973) pp. 245-256.

2. J. R. Rairden, Proc. 9th Int. Thermal Spraying Conf., Nederlands Instituut voor Lastechniek, The Hague, Netherlands (1980) pp. 329-333.

3. M. R. Jackson, J. R. Rairden, J. S. Smith and R. W. Smith, J. of Metals 33 (1981) pp. 23-26.

4. J. R. Rairden, M. R. Jackson, M. F. X. Gigliotti, M. F. Henry, J. R. Ross, W. A. Seaman, D. A. Woodford and S. W. Yang, General Electric Report No. 81CRD075, Schenectady, New York (1981).

5. F. J. Pennisi and D. K. Gupta, Thin Solid Films 84 (1981) pp. 49-58.

6. F.J. Hermanek, Int. J. Powder Met. Powder Tech. 18 (1982) pp. 81-85.

7. H. C. Fiedler, Plasma Processing and Synthesis of Materials, J. Szekely and D. Apelian, eds., Elsevier, New York, NY (1984) pp. 173-180.

8. N. R. Shankar, H. Herman and R. K. MacCrone, Plasma Processing and Synthesis of Materials, J. Szekely and D. Apelian, eds., Elsevier, New York, NY (1984) pp. 181-189.

9. D. R. Mash and I. Mac P. Brown, Met. Eng. Quart., Feb. (1964) pp. 18-25.

10. A. R. Moss and W. J. Young, Powder Met. 7 (1964) pp. 261-289.

11. P. A. Siemers, M. R. Jackson, R. L. Mehan and J. R. Rairden, General Electric Report No. 85CRD001, Schenectady, NY (1985).

12. M. O. Price, T. A. Wolfla and R. C. Tucker, Thin Solid Films 45 (1977) pp. 309-319.

13. M. R. Jackson, P. A. Siemers, J. R. Rairden, R. L. Mehan and A. M. Ritter, this proceedings.

14. M. R. Jackson and A. M. Ritter, "Chemical and Structural Compatibility in γ/γ'-α Composites", General Electric Report No. 86CRD250, Schenectady, NY (1987).

15. B. I. Edelson and W. M. Baldwin, Trans. ASM 55 (1962) pp. 230-250.

16. J. Gurland and N. M. Parikh, Fracture, Vol. 7, H. Liebowitz, ed., Academic Press, New York, NY (1972) pp. 841-878.

17. E. Orowan, Symp. Internal Stresses in Metals and Alloys, Institute of Metals, London (1948) pp. 451-460.

18. M. F. Ashby, Phil. Mag. 14 (1966) pp. 1157-1178.

19. M. Gensamer, E. B. Pearsall, W. S. Pellini and J. R. Low, Trans. ASM 30 (1942) pp. 983-1020.

20. T. B. Cox and J. R. Low, Metall. Trans. 5 (1974) pp. 1457-1459.

21. T. Gladman, B. Holmes and I. D. McIvor, Effects of Second-Phase Particles onthe Mechanical Properties of Steels, Iron and Steel Institute, London (1971) pp. 88-99.

22. B. J. Brindley and T. C. Lindley, JISI 211 (1972) pp. 124-125.

23. T. L. Cheeks, M. E. Glicksman, M. R. Jackson and E. L. HAll, General Electric Report No. 83CRD262, Schenectady, New York (1983).

24. R. C. Ruhl, Mat. Sci. Eng. 1 (1967) pp. 313-320.

25. H. Jones, Proc. Int. Conf. Rapid Sol. Processing, R. Mehrabian, B. H. Kear, and M. Cohen, eds. Claitor's Pub. Division, Baton Rouge, LA (1977) pp. 28-45.

26. J. Gurland, Trans. AIME 227 (1963) pp. 1146-1150.

27. M. F. Ashby, Oxide Dispersion Strengthening, Proc. 2nd Bolton Landing Conf., G. S. Ansell, T. D. Cooper and F. V. Lenel, eds., Gordon and Breach, New York, NY (1966) pp. 143-211.

28. R. H. Jones, Metall, Trans. 4 (1973) pp. 2799-2808.

29. R. F. Decker, Metall. Trans. 4 (1973) pp. 2495-2512.

30. P. M. Kelly, Scripta Met. 6 (1972) pp. 647-656.

31. S. N. Singh and M. C. Flemings, Trans. AIME 245 (1969) pp. 1811-1819.

32. A. Gangulee and J. Gurland, Trans. AIME 239 (1967) pp. 269-272.

33. R. H. Van Stone and T. B. Cox, ASTM STP-600, American Society for Testing and Materials, Philadelphia, PA (1976) pp. 5-29.

34. A. M. Ritter, PhD Thesis, Rensselar Polytechnic Institute, Troy, NY (1987).

35. Y. S. Touloukian, R. K. Kirby, R. E. Taylor, and T. Y. R. Lee, Thermophysical Properties of Matter, Vol. 13, Plenum Press, New York, NY (1972).

36. H. T. Corten, Modern Composite Materials, L. J. Broutman and R. H. Krock, eds., Addision-Wesley, Menlo Park, CA (1967) p. 27.

37. M. F. Henry, PhD Thesis, Rensselaer Polytechnic Institute, Troy, NY (1974).

CREEP DEFORMATION OF WC-CO ALLOYS

Gwendolyn Dixon and Roger N. Wright

Materials Engineering Department
Rensselaer Polytechnic Institute
Troy, New York

Abstract

WC-Co alloys with a grain size of 1.1 μm and WC volume fractions of 68% and 76% have been evaluated by stress relaxation test methods. Stress-strain rate-temperature relations have been developed, temperature coefficients have been determined, and a simple power law relation between stress and strain rate has been quantified. Pure cobalt has been similarly evaluated. In the 950-1000 °C range, stress and strain rate relations for WC-Co flow are somewhat consistent with cobalt matrix flow mechanisms. At lower test temperatures, flow stresses, temperature coefficients, and stress exponent (1/m) values are much too high to be associated with unconstrained cobalt flow.

Processing and Properties for Powder Metallurgy Composites
Edited by P. Kumar, K. Vedula and A. Ritter
The Metallurgical Society, 1988

Introduction

WC-Co alloys have very high hardness and good wear resistance and, as such, are primarily used for metal cutting purposes. Other uses include rock drilling, stone cutting, metal forming tools, wear parts and abrasive grits. These applications often involve extended exposure to elevated temperature and stress, and the potential for plastic deformation must be considered. A number of high temperature deformation studies of WC-Co (cemented carbides) have been published, including data from conventional creep tests, compression tests, and stress relaxation tests. Figure 1 shows data compiled from the work of Wirmark and Dunlop (1), Smith and Wood (2), Schenck, et al. (3), Doi, et al. (4), and Osterstock (5). In the 750 to 1200 °C regime, flow stresses range from below 50 to over 700 MPa, over a strain rate range from below 10^{-8} to nearly 10^{-3} s^{-1}.

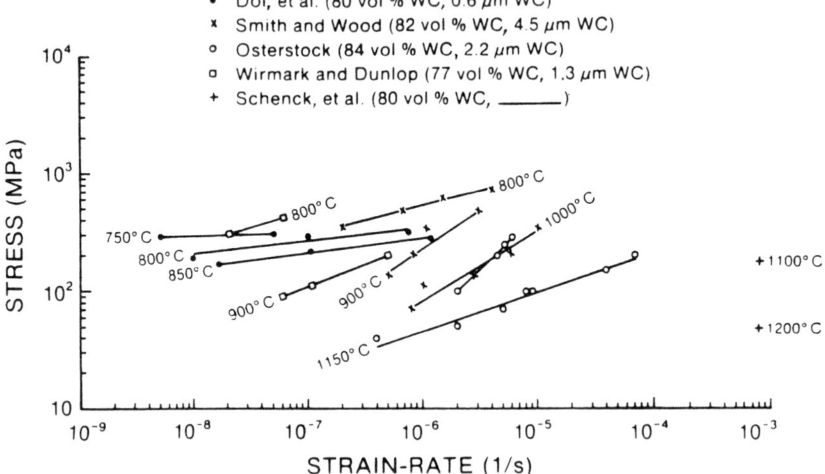

Figure 1. Creep data compiled from Wirmark and Dunlop (1), Smith and Wood (2), Schenck, et al. (3), Doi, et al. (4), and Osterstock (5).

A number of deformation mechanisms have been proposed to account for the data in Figure 1. Smith and Wood (2) proposed that, at low stresses, diffusion of Co around WC grains is the predominant creep mechanism, whereas at high stresses the principal mechanism involves dislocation looping between WC grains. Wirmark and Dunlop (1) surmised that sliding of WC grain boundaries and WC grain rotation were the major factors in elevated temperature creep of WC-Co.

The present study makes extensive use of stress relaxation technique to generate and interpret stress-strain rate-temperature relations in a range germane to the modeling of high speed metal cutting.

Procedure and Results

Materials

Two WC-Co alloys were studied, namely Carboloy grades 779 and 55A, referred to hereafter as grades A and B, respectively. Carboloy is a registered tradename of the General Electric Company and that company provided the alloys for this research, in the form of 4 mm diameter rods. Figures 2 and 3 show the as-received microstructure for grades A and B. Table I lists the average WC grain diameter, WC volume fraction, contiguity, and mean free path. Contiguity is defined as the degree of contact between particles of a given phase in a composite microstructure. The contiguity is influenced by sintering conditions and is strongly dependent on the binder content (6) and on the dihedral angle created by contacting grains of the given phase (7). The mean free path between WC grains is a function of cobalt content and WC grain size. A detailed consideration of the quantitative metallographic measurements has been set forth by Dixon (8).

Table I. Quantitative Metallographic Description of Grades A and B

WC-Co Grade	Average WC Grain Diameter	WC Volume Fraction	Contiguity of WC	Mean Free Path in Cobalt
A	1.12 μm	0.76	0.51	0.67 μm
B	1.12	0.68	0.44	0.78

Figure 2. Polished and etched (Murakami's reagent) microstructure of Grade A WC-Co. This as-received microstructure shows angular WC grains embedded in a cobalt matrix, with the WC grains exhibiting a wide range of sizes.

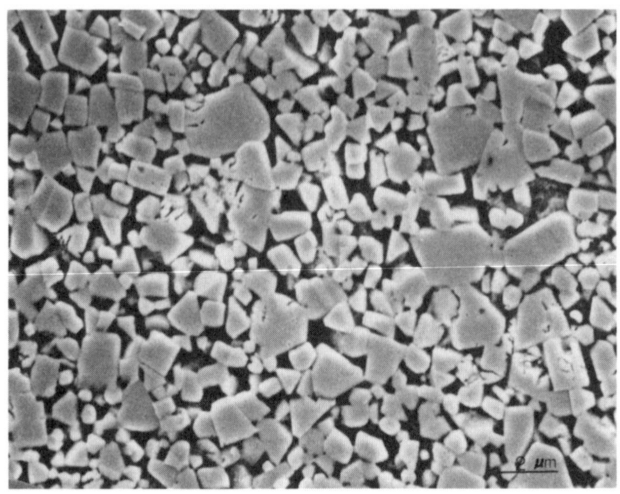

Figure 3. Polished and etched (Murakami's reagent) microstructure of grade B WC-Co.

The behavior of pure cobalt was observed as a basis for interpreting the WC-Co results. Cobalt was purchased from Alfa Products in the form of 5mm diameter, cold worked rods. The purity level was 99.8%, and the as-received hardness was DPH 300. The cobalt was conditioned prior to stress relaxation testing by annealing in an evacuated pyrex capsule for 55 hours at 475 °C, followed by quenching in ice-water according to the procedure of Bibby and Parr (9). The annealing process was performed to ensure that the initial cobalt structure would be face-centered-cubic. The initial cold worked structure is shown in Figure 4, and the annealed structure is shown in Figure 5. The 475 °C anneal does not produce a recrystallized structure. To assess microstructural evolution in the testing temperature range, a series of 475 °C annealed samples was additionally annealed for two hours at 800, 900, 950, and 1000 °C. As shown in Figures 6 through 9, these higher temperature anneals produce some structural coarsening with grain sizes roughly in the 8 µm range.

Theory of Stress Relaxation Test Methods

The stress relaxation or load relaxation test involves monitoring the reduction of stress with time in a specimen that has been loaded and held under constant constraint. Typical output from a load relaxation test is displayed in Figure 10. The specimen is loaded at a constant crosshead rate to some predetermined load level, point A in Figure 10, and all crosshead motion stopped. The specimen continues to deform plastically, whereas the load train and specimen unload elastically. The time rate of change of the applied load is a direct measure of the plastic deformation rate of the specimen. There are various advantages to using a stress relaxation technique to study stress-strain rate-temperature interactions in WC-Co alloys. In each stress relaxation test, the sample normally only experiences a small increment of plastic strain. Therefore, there is little strain-related microstructural evolution, and multiple tests can often be run on the same sample with little structural change. Many problems presented by compression testing, such as plastic buckling and barreling, are largely avoided by stress relaxation testing. Stress relaxation testing subjects the specimen to continuous reductions of load and deformation rate, and therefore a spectrum of stress and strain-rate values is provided by a single test.

Figure 4. As-received cobalt microstructure, prior to annealing.

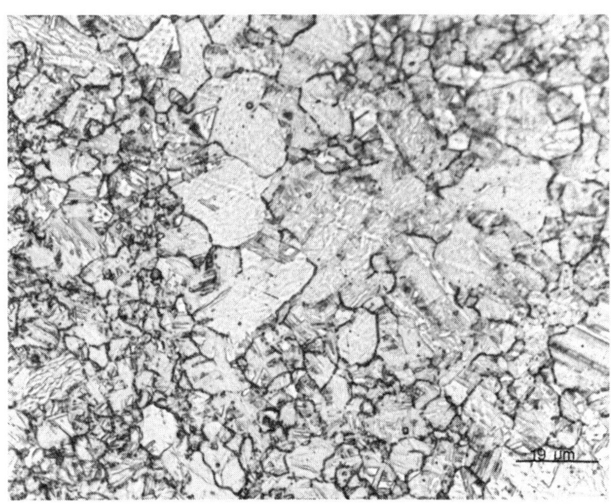

Figure 5. Cobalt microstructure after annealing 55 hours at 475 °C.

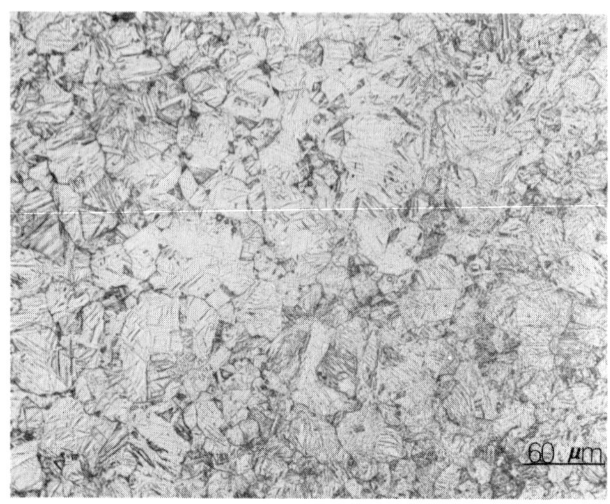

Figure 6. Cobalt microstructure after annealing for 2 hours at 800 °C.

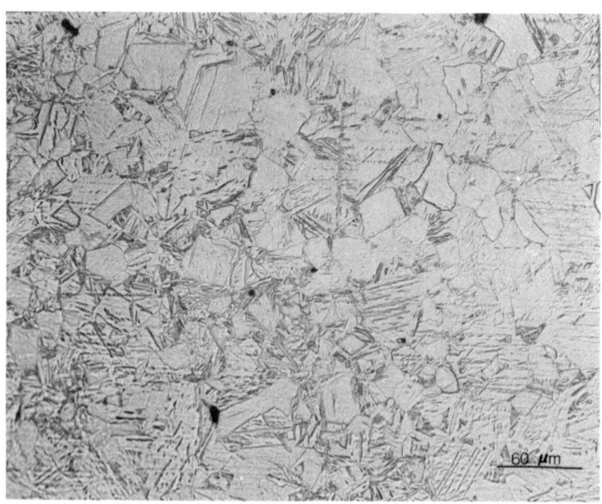

Figure 7. Cobalt microstructure after annealing for 2 hours at 900 °C.

Figure 8. Cobalt microstructure after annealing 2 hours at 950 °C.

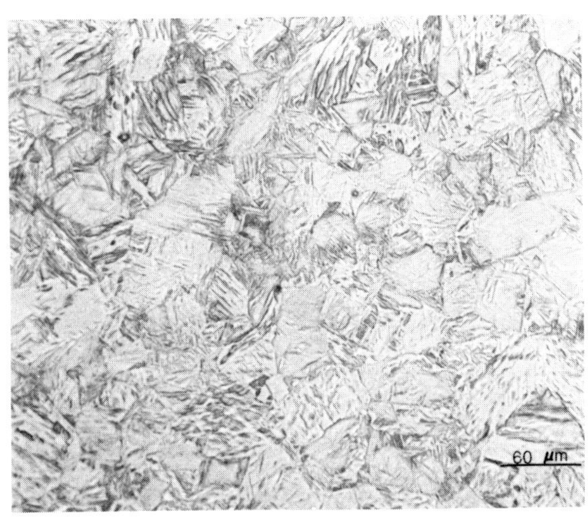

Figure 9. Cobalt microstructure after annealing for 2 hours at 1000 °C.

Load Relaxation Data Analysis

As shown in Figure 10, the output from a load relaxation test gives load as a function of time. To convert this data to stress-strain rate relationships, the output is divided into constant load decrements. An average stress for a given decrement is obtained by dividing the average load by the initial cross sectional area. The average strain rate for the decrement is obtained from the following equation:

$$d\varepsilon/dt \;=\; dP/dt\,[1/(EA_0) + 1/(KL_0)]\;, \qquad (1)$$

where $d\varepsilon/dt$ is strain rate, dP/dt is the load decrement divided by the time taken for the load decrement, E is Young's modulus for the specimen, A_0 is the initial cross sectional area of the sample, K is the testing machine stiffness, and L_0 is the initial specimen length. The strain rate that is defined by equation (1) can include anelastic relaxations as well as plastic deformation. The present study has not considered anelastic response, and has employed a simple model converting elastic strain into plastic strain.

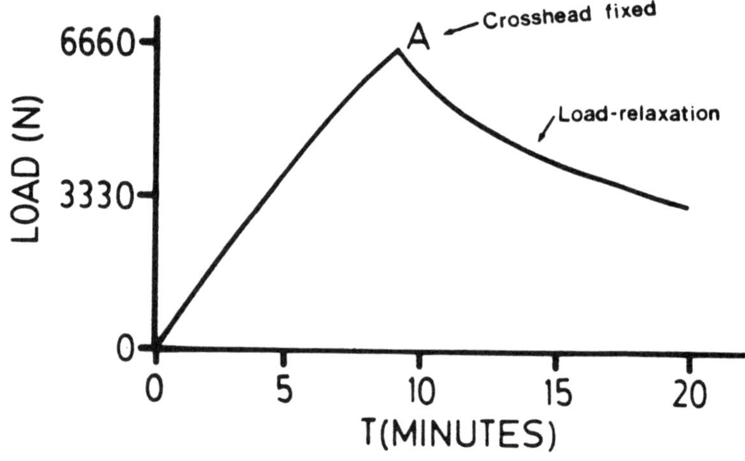

Figure 10. Typical load relaxation test output.

Experimental Apparatus

Figure 11 shows a schematic of the experimental set-up used for the load relaxation testing. All samples were tested in a compressive mode, and the compression chamber was mounted on a 44,500 N capacity Instron testing machine. Inside the compression chamber, the specimens were mounted between two alumina heads, with alumina being used to minimize chemical interaction with the test specimens. Test temperatures were achieved through induction heating. Temperature control was within 10 °C of the nominal test temperature. The compression chamber was enclosed in an opaque quartz tube. The test environment was flowing argon gas. Prior to heating and testing, a small load was placed on the specimen to stabilize the load train. Frequent downward adjustment of this load was necessary during heating because of thermal expansion. At the desired temperature, the sample was loaded to the desired load level, and all crosshead motion stopped. The load relaxation response was then monitored as a function of time.

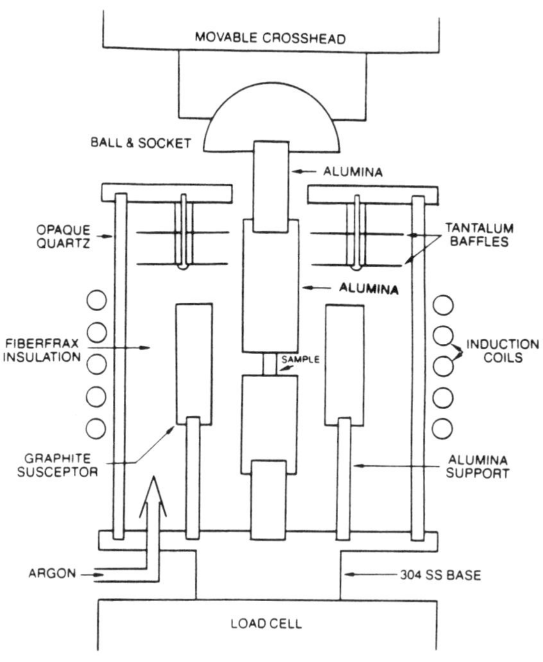

Figure 11. Schematic representation of experimental apparatus.

Stress Relaxation Testing of WC-Co

Stress relaxation tests were performed at 800, 900, 950, and 1000 °C. Specimens were cut from the rod stock with length-to-diameter ratios of about 2.5. Each specimen was loaded at a constant crosshead rate of 0.85 μm/s to a starting load level of 6670 N. The stress-strain rate relations calculated from the stress relaxation test data (see Load Relaxation Data Analysis above) are shown in Figures 12 and 13. The single line at a given temperature represents linear regression analysis of data consolidated from six separate tests. These log stress versus log strain rate plots reveal that the slopes for each alloy increase as temperature increases. These slopes give the value of the strain rate sensitivity exponent, m (or 1/n where n is a stress exponent). The strain rate sensitivity exponents derived from the data in Figures 12 and 13 are presented in Table II. The implications are that strain rate sensitivity increases with increases in temperature, with increase in cobalt content, with increase in mean free path in the cobalt, and with decrease in WC contiguity.

Table II. Strain Rate Sensitivity as a Function of Temperature for WC-Co and Cobalt

Material	800 °C	900 °C	950 °C	1000 °C
Grade A	0.01	0.04	0.10	0.29
Grade B	0.02	0.08	0.21	0.39
Cobalt	0.15	0.10	0.11	0.12

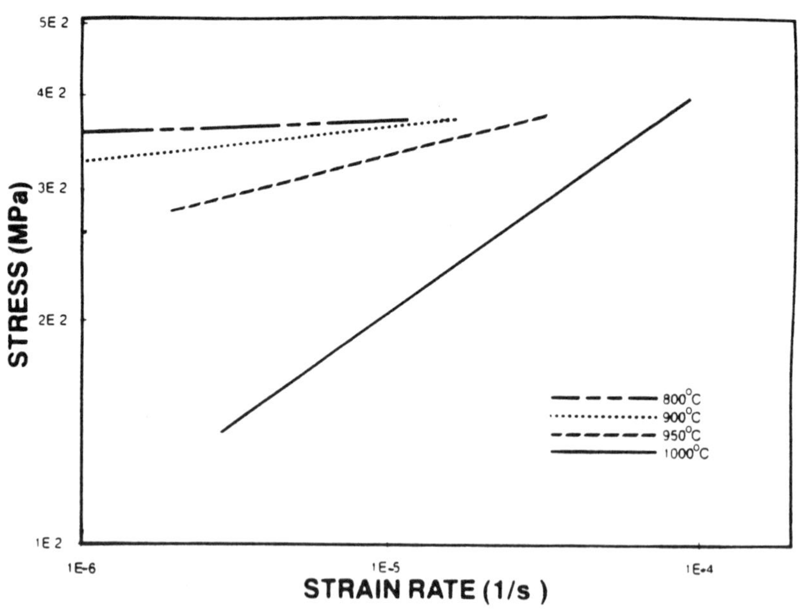

Figure 12. Stress-strain rate data for Grade A WC-Co.

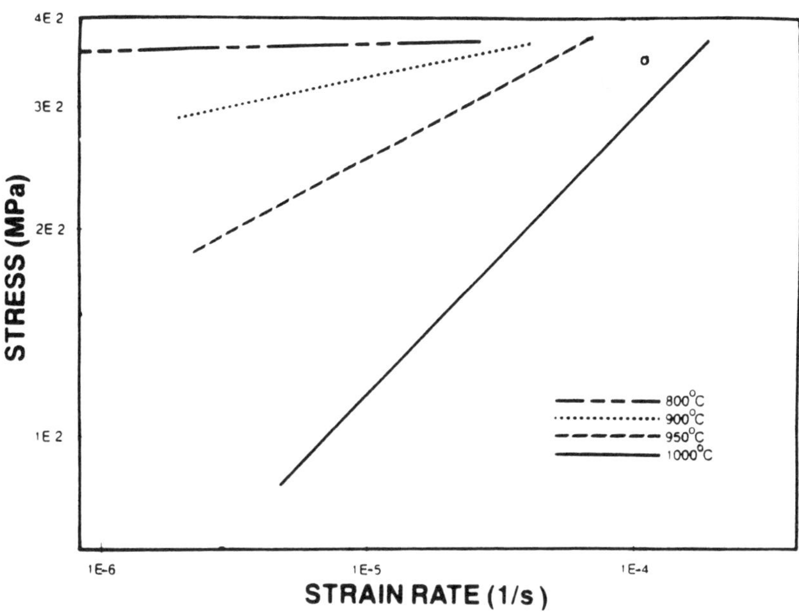

Figure 13. Stress-strain rate data for Grade B WC-Co. Circle denotes compression test result.

Stress-Strain Compression Tests

Monotonic stress-strain testing was undertaken in compression for Grade B at 1000 °C to compare such data with the relations derived from stress relaxation testing. The results are shown in Figure 14. The cylindrical specimen had a length-to-diameter ratio of 2.0. The test was run at a crosshead speed of 0.85 µm/s in an argon atmosphere. With a steady state flow stress of 375 MPa at a strain rate of $1.4 \times 10^{-4}\ s^{-1}$, fair agreement exists with the relationships of Figure 13. The lower plastic strain rate estimates for the lower strains of Figure 14 reflect the elastic displacements of the testing machine.

Plastically Deformed WC-Co Microstructure

After plastic strains of several percent from monotonic compression or repeated stress relaxation testing, porosity development was observable at WC-Co interfaces. This is shown in Figures 15 and 16 for a Grade B specimen which had undergone cumulative plastic strain of 0.11 after repeated stress relaxation testing. Some porosity was present, a priori, as may be seen in the micrograph of the as-received structure shown in Figure 17. Porosity development was not considered to have a major influence on the observed stress-strain rate-temperature relationships, particularly in view of the reproducible results achievable with repeated stress relaxation testing of a single specimen.

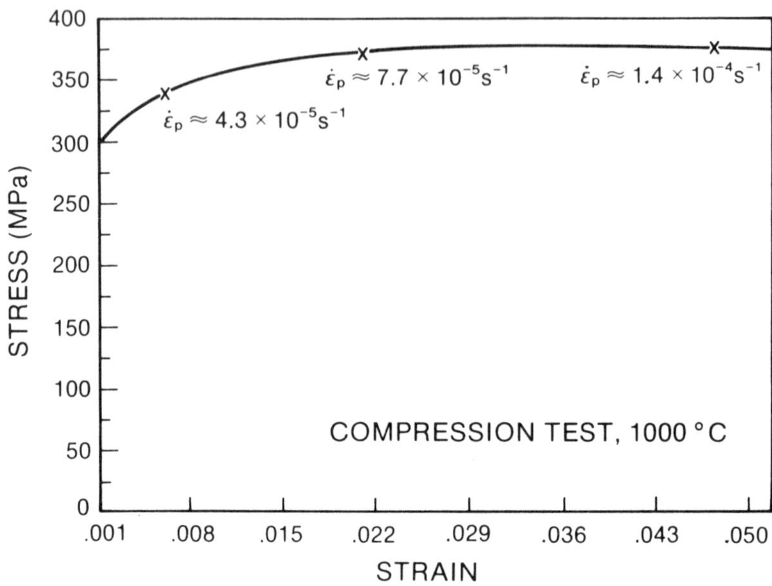

Figure 14. True stress-true strain curve for grade B WC-Co at 1000 °C.

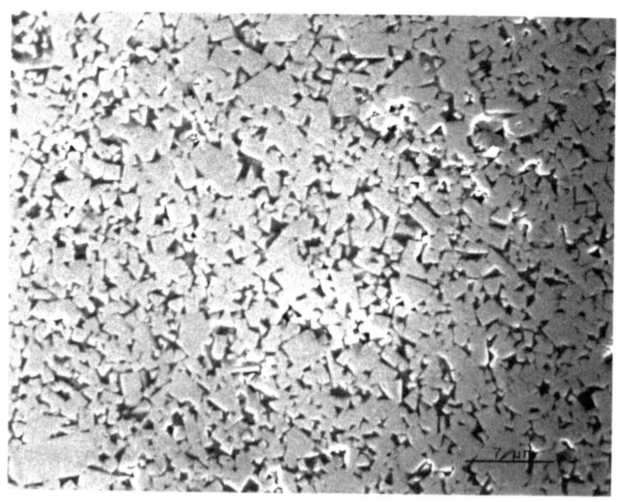

Figure 15. Transverse section of Grade B stress relaxation specimen, taken at the centerline. The accumulated plastic strain is 0.11.

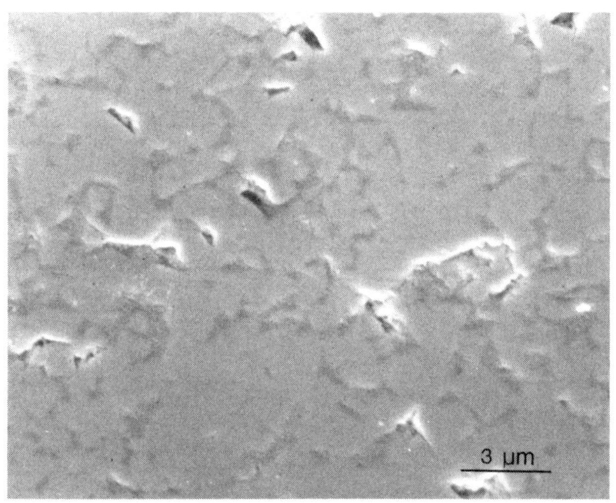

Figure 16. Higher magnification view of structure in Figure 15.

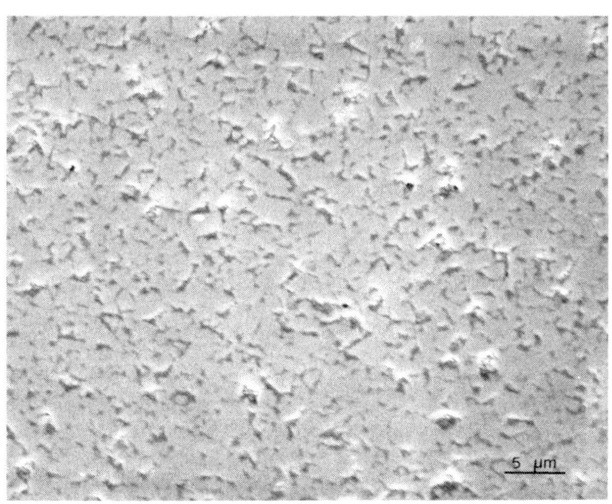

Figure 17. As-received microstructure of Grade B, showing a priori porosity.

Stress Relaxation Testing of Cobalt

Figure 18 shows the stress-strain rate relations derived from the stress relaxation testing results for the cobalt specimens tested under the same conditions as the WC-Co. Again, the single line at a given temperature represents the consolidation of six separate tests. The strain rate sensitivity values for the cobalt are shown in Table II.

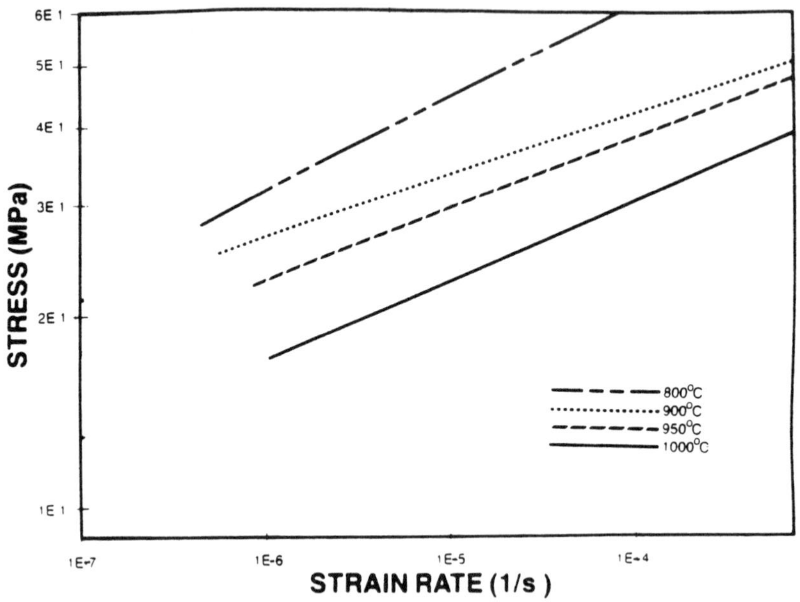

Figure 18. Stress-strain rate data for cobalt.

Temperature Coefficients

Creep is a thermally activated process and can be described by a rate equation (10), as follows:

$$d\varepsilon/dt = C \exp(-B/RT), \qquad (2)$$

where B is a temperature coefficient for the rate controlling process, C is a pre-exponential constant, T is absolute temperature, and R is the universal gas constant. By assuming that C remains constant during small temperature changes, B can be calculated as follows:

$$B = R [\ln(\dot{\varepsilon}_1/\dot{\varepsilon}_2) T_1T_2/(T_1-T_2)], \qquad (3)$$

where $\dot{\varepsilon}_1$ and $\dot{\varepsilon}_2$ are, respectively, the strain rates for a given stress at temperatures T_1 and T_2. Temperature coefficients derived on this basis are given in Table III. At 300 MPa, the temperature coefficients for grades A and B range from about 400 to 1200 kJ/mole. The temperature coefficients decrease with increasing temperature. Temperature coefficients for cobalt at a stress of 30 MPa are also given in Table III. The temperature coefficients for cobalt range from 200 to 500 kJ/mole and increase with increasing temperature. The temperature coefficients in Table III are cited with uncertainties representing one standard deviation.

Discussion

Stress-Strain Rate Relations for WC-Co

The stress relaxation data were plotted as log stress versus log strain rate for the WC-Co grades and for cobalt. The empirical relationship assumed by such data analysis is expressed by the following general relationship between flow stress and strain rate:

$$\sigma = S\,(d\varepsilon/dt)^m, \tag{4}$$

where σ is flow stress and S is a proportionality constant depending on temperature and microstructure. This relationship is useful in modeling WC-Co responses in manufacturing processes, and values for S and m are set forth in Table IV, together with microstructural information. The flow stress appears to increase with increasing WC volume fraction, increasing WC contiguity, and decreasing cobalt mean free path.

Table III. Temperature Coefficients

Material	Stress	Temperature Range	B Value
Grade A	300 MPa	800 - 900 °C	1050 ± 40
		900 - 950	870 ± 300
		950 - 1000	590 ± 10
Grade B	300	800 - 900	520 ± 130
		900 - 950	395 ± 5
		950 - 1000	380 ± 80
Cobalt	30	800 - 900	160 ± 50
		900 - 950	320 ± 25
		950 - 1000	500 ± 230

Stress-Strain Rate Behavior of Cobalt

The values of S and m for the case of the cobalt data are given in Table V, along with values taken from the creep data of Sargent, Malakondaiah, and Ashby (11). The large difference in strain rate sensitivity values between the current results and those of Sargent, et al. can be attributed to the essentially constant structure of the current stress relaxation testing. As Sherby, et al. (12) have discussed for pure metals, creep under constant structure conditions can be expected to display a stress exponent (1/m) in the range of eight, whereas a stress exponent of five is expected under creep conditions allowing substructural coarseness to adjust to stress level. The stress relaxation data for cobalt allow a more rigorous analysis of the WC-Co creep data.

Table IV. Evaluation of S and m from Equation (4) for WC-Co (1/m is given in parenthesis)

Grade	f_{WC}	Contiguity	Mean Free Path (μm)		1000°C	950°C	900°C	800°C
A	75.8	0.51	0.67	**S**	$6x10^3$	$1x10^3$	590	436
				m	0.29	0.10	0.04	0.01
					(3.4)	(10.0)	(23.2)	(71.4)
B	67.7	0.44	0.78	**S**	$11x10^3$	$2.88x10^3$	863	436
				m	0.39	0.21	0.08	0.02
					(2.5)	(4.7)	(12.0)	(66.7)

Table V. Evaluation of S and m from Equation (4) for WC-Co (1/m is given in parenthesis)

Cobalt		1000°C	950°C	900°C	800°C
	S	94.6	105.7	100.7	240.9
	m	0.12	0.11	0.09	0.15
		(8.1)	(8.9)	(10.3)	(6.7)
*	**S**	571	726	760	483
	m	0.29	0.29	0.28	0.19
		(3.36)	(3.4)	(3.6)	(5.1)

*From Sargent, Malakondaiah, and Ashby

Analysis of WC-Co Creep Behavior

Smith and Wood (2) suggested that high stress creep of WC-Co alloys is consistent with a model developed by Ansell and Weertman (13) for dispersion strengthened alloys. This model proposes that for a given particle separation and a given volume fraction of second phase particles, there are minimum stresses which must be exceeded to move dislocations through the matrix and between second-phase particles. The matrix dislocations should loop between the dispersed particles, and creep should occur as the dislocation loops climb away to allow a second matrix dislocation to pass between the particles. The Ansell and Weertman model involves the following behavior:

1. The creep rate increases with temperature;

2. The creep rate increases with the fourth power of stress ($m = \frac{1}{4}$);

3. At constant stress, the creep rate increases with the square of the interparticle spacing;

4. The activation energy is equivalent to the activation energy for self-diffusion in the matrix.

The present data shows some consistency with this model at the highest testing temperatures. The temperature coefficients for WC-Co are roughly comparable with those for cobalt in the 950-1000 °C range, whereas at lower temperatures a large discrepancy opens up as the WC-Co temperature coefficients increase while those of cobalt decrease. On the other hand, the cobalt temperature coefficients are, in any case, much higner than the activation energy for cobalt self-diffusion. At 950-1000 °C, the m values for WC-Co are near the value projected by the Ansell and Weertman model. The low m values found in the lower temperature ranges are indicative of a mechanism involving a much stronger influence of stress on the deformation process.

Thus, at the highest test temperatures, stress-strain rate relations are somewhat consistent with cobalt matrix flow. At the lower test temperatures, flow stresses, temperature coefficients, and stress exponent (1/m) values are much too high to be associated with unconstrained cobalt flow. The constraint and intrinsic strength of the WC "skeleton" may be dominant, and deformation mechanisms may include WC grain boundary sliding and grain rotation, as suggested by Wirmark and Dunlop (1).

Summary

Constitutive equations have been developed relating stress, temperature, strain rate, and microstructure. In particular, power-law relations between stress and strain rate have been quantified for the 800-1000 °C range. Using the simple power-law relation

$$\sigma = S\,(d\varepsilon/dt)^m,$$

both the proportionality constant, S, and the strain rate sensitivity, m, have been shown to increase with temperature in the test temperature range. Using the relation

$$d\varepsilon/dt = C\exp(-B/RT),$$

The temperature coefficient, B, is seen to decrease with increasing temperature. In the 950-1000 °C range, stress and strain rate relations for WC-Co flow are somewhat consistent with cobalt matrix flow mechanisms. At lower test temperatures, flow stresses, temperature coefficients, and stress exponent (1/m) values are much too high to be associated with unconstrained cobalt flow.

Acknowledgments

The authors wish to acknowledge the support of the Forging Industry Association and Rensselaer Polytechnic Institute. Dr. Dixon was supported by a Graduate Professional Opportunity Program Fellowship during this work. The authors wish to extend special thanks to Dr. Minyoung Lee of the General Electric Corporate Research and Development Center for generously supplying specimen materials and experimental guidance.

References

1. G. Wirmark and G. L. Dunlop, Proceedings of the International Conference on the Science of Hard Materials, 1984, ed. E. A. Almond, C. A. Brooks, and R. Warren (Accord, MA: Institute of Physics Conference Series Number 75, Adam Hilger Ltd., 1986), 669.

2. J. T. Smith and J. D. Wood, Acta Metallurgica, 16 (1968), 1219.

3. S. R. Schenck, R. J. Gottschall, and W. S. Williams, Materials Science and Engineering, 32 (1978), 229.

4. H. Doi, F. Ueda, Y. Fujiwara, and H. Masatomi, International Journal of Refractory and Hard Metals, 3 (3) (1984), 146.

5. F. Osterstock, Science of Hard Materials, ed. R. Viswananadham, D. Rowcliffe, and J. Gurland (New York, NY: Plenum Pub. Corp., 1983), 671.

6. J. Gurland, Transactions TMS-AIME, 212 (1958), 422.

7. R. M. German, Metallurgical Transactions A, 16A (1985), 1247.

8. G. Dixon, "Mechanical Behavior of Cemented Carbides" (Ph.D. thesis, Rensselaer Polytechnic Institute, 1987), 52.

9. M. J. Bibby and J. G. Parr, Cobalt, 20 (1963), 111.

10. J. E. Dorn, Creep and Recovery (Metals Park, OH: American Society for Metals, 1957), 255.

11. P. M. Sargent, G. Malakondaiah, and M. F. Ashby, Scripta Metallurgica, 17 (1983), 625.

12. O. D. Sherby, R. H. Klundt, and A. K. Miller, Metallurgical Transactions A, 8A (1977), 843.

13. G. S. Ansell and J. Weertman, Transactions TMS-AIME, 215, (1959), 838.

CHARACTERISTICS OF P/M PROCESSED Cu-Nb COMPOSITES

C. L. Trybus, W. A. Spitzig, J. D. Verhoeven, and F. A. Schmidt

Ames Laboratory-USDOE
Iowa State University
Ames, IA 50011

Abstract

The feasibility of producing high strength *in situ* Cu-Nb composites via P/M processing techniques is explored because P/M offers a much wider range of compositions for the composites as compared to traditional ingot metallurgy. In this study two Cu-20 vol.% Nb composites were produced by hot extrusion of encapsulated Cu and Nb powders. Both composites were subsequently swaged into wire and mechanical properties were evaluated as a function of the amount of deformation. One composite comprised of spherical elemental particles did not display increased strength upon deformation while another one made from irregularly shaped powders did possess increased strength. The strength-deformation strain relationships are correlated with microstructural characteristics. These results are compared to the properties exhibited by cast Cu-Nb *in situ* composites.

Processing and Properties for Powder Metallurgy Composites
Edited by P. Kumar, K. Vedula and A. Ritter
The Metallurgical Society, 1988

Introduction

Cu-Nb alloys can be formed into *in situ* composites. These alloys consist of a mixture of Cu and Nb from which the composite is formed upon mechanical deformation. The volume fraction of Nb is kept low (less than 20 volume per cent) so that the Nb is essentially isolated within the Cu matrix. Rolling, wire drawing, and swaging can each cause the Nb to attain a filamentary morphology parallel to the direction of working.

Tensile strengths reported for *in situ* Cu-Nb composites are much greater than those predicted by the rule of mixtures [1-2]. The high strengths of these materials have been attributed to the ribbon-like cross sections developed by the Nb filaments [1]. Upon wire drawing of a pure BCC material, like Nb, the <110> directions align themselves parallel to the wire axis which results in Nb undergoing a plane-strain mode of deformation [3]. In the axisymmetrical deforming composite, this causes the Nb filaments to kink and fold about the wire axis in order to maintain compatibility between them and the Cu matrix. Thus, the Nb filament morphology can be described as long flat ribbons with highly irregular cross sections.

Recently, Spitzig et al. [4] showed that the strength of heavily cold worked wires of Cu-Nb increased as the spacing between Nb filaments decreased. It was found that the composite obeyed a Hall-Petch type relationship indicating that the Nb filaments act as planar barriers to dislocation motion between the Cu and Nb. Clearly, the unique microstructures developed in *in situ* Cu-Nb composites are the source of the strength increases observed.

The majority of Cu-Nb composites have been fabricated from billets formed by two methods, casting and powder metallurgical (P/M) techniques. Casting methods including consumable arc melting [5] and rf levitation melting [1], have been utilized to produce high strength *in situ* Cu-Nb composites. P/M methods appear to be favored for the production of wire to be used in superconductor applications. Both hot [6-9] and cold [9-13] extrusion of loose powders have been employed to produce Cu-Nb composites. These composites are then reacted with Sn to form the superconducting phase, Nb_3Sn. Although P/M billets have been successfully used to produce superconducting wires, they have not been proven suitable for the fabrication of high strength *in situ* Cu-Nb composites. The goal of this research was to fabricate high strength Cu-Nb composites via P/M. The microstructural characteristics and tensile strengths were evaluated and compared to consumable arc melted composites.

Experimental Procedures

Two billets comprised of elemental Cu and Nb powders were hot extruded at 750°C at an areal reduction of 12:1. A third compact made from powders prealloyed by REP (rotating electrode process) was also hot extruded at 750°C at 14:1 areal reduction. The properties of the Cu-Nb REP powder and the resulting composite microstructure are given elsewhere [14]. Composites made from REP powder did not show increases in strength with increasing cold working. The characteristics of the elemental powders employed are given in Table 1 and Figure 1. The Nb must be ductile in order for it to become filamentary during cold work. Nb powder produced by the hydride-dehydride process has been successfully employed in P/M processing of superconducting Cu-Nb-Sn composites [13].

In all three cases the powders were mixed and packed to tap density into an extrusion can of 101 copper 7.62 cm in diameter. Prior to mixing, the elemental Cu powders were heated in flowing H_2 at $\sim$ 425°C for 4 hours

Table 1. Initial Powder Characteristics.

Extrusion No.	Powder Size Range (μm) Cu	Nb	Other
1	200-400	200-400	Spherical Cu shot Spherical Nb REP
2	< 44	88-125	Irregular Cu Produced from reduced oxide. Hydride-dehydride Nb (Fig. 1).

to reduce surface oxides. The extrusion can top, which was shaped as a truncated cone, had a concentric series of 30 holes (1.5 mm nominal diameter) drilled parallel to the longitudinal axis of the can half way up the conical surface. After the can was packed with powder it was placed under vacuum in an electron beam welder and heated to approximately 800°C by a defocused electron beam. This heating cycle was repeated 3 times over a 24 hour period. Subsequently, the holes in the cap were welded shut. This procedure was used to remove absorbed gases from the powder surface.

Post extrusion forming processes included swaging from 2.22 cm to 1.27 cm diameter, removing of the copper clad, followed by swaging end-over-end (reversing the direction of swaging after each pass) to a diameter 0.64 mm. Successively smaller wires were drawn to a minimum wire diameter of 0.15 mm. This minimum diameter represents the smallest diameter wire which we are able to test accurately and is not indicative of composite fracture during drawing. The level of deformation will be given in logarithmic strain by $\eta = \ln (A_0/A)$, termed the draw ratio, where A_0 and A are the initial and final cross sectional areas, respectively. In this study the largest draw ratio attained was $\eta = 9.4$ which corresponds to a reduction in area of 99.992%.

Tensile specimens were machined from the extruded and swaged material and from wires drawn down to 0.25 cm in diameter. Smaller wires were tested by embedding their ends into Cu sleeves using a superbond adhesive. All tensile tests were done at 22°C using a nominal strain rate of 1.7×10^{-4} $sec.^{-1}$ Multiple specimens (2 to 3) were tested at each wire size and the results were the same, within 5%.

Fig. 1. Scanning electron micrograph of the Nb hydride dehydride powder used in extrusion 2.

Samples for microstructural examination were prepared using a standard series of fine grinding steps (SiC) followed by final polishing in alumina slurries. All samples were lightly etched in a solution of $1H_3PO_4:1H_2O_2:1.7H_2O$. Filament sizes and spacings were measured using a metallograph on the longitudinal sections. Standard quantitative metallographic techniques were employed in these calculations [15].

Results and Discussion

The as-extruded microstructures of extrusions 1 and 2 are shown in Figures 2 and 3. Pores at the Nb and Cu interface (Fig. 2) enlarged upon swaging and no increase in strength was observed in this material due to their persistent presence. Mechanical test results on extrusion 2 are given in Figure 4. For comparison, results are also shown for wire drawn pure Cu and for a cast composite [4]. Both the cast and extruded composite show large increases in ultimate tensile strength over pure Cu at draw ratios greater than ~ 6. At low draw ratios (<3) the P/M composite is stronger than the cast material due to an additional increment in strength from the refined Cu matrix grain size. The grain size of Cu in the as-extruded material (Fig. 3) is about half that of the as-cast. In addition, it is difficult to compare the as-extruded composite directly with the as-cast material. During the extrusion process the composite is mechanically worked while in casting no such deformation occurs. As the deformation level increases both of these effects are eliminated.

Interstitially dissolved oxygen in Nb has a deleterious effect on its post-extrusion deformability [6]. Borman et al. [6] reported oxygen levels below 0.1 at.% in the Nb after extrusion of a Cu-Nb powder mixture. In these studies the initial Nb powder had more than twice as much oxygen (Table 2) and yielded a well developed Nb filamentary structure upon cold working. The oxygen content in extrusion 1 was lower than in 2, indicating that excessive oxygen was not the cause of its failure.

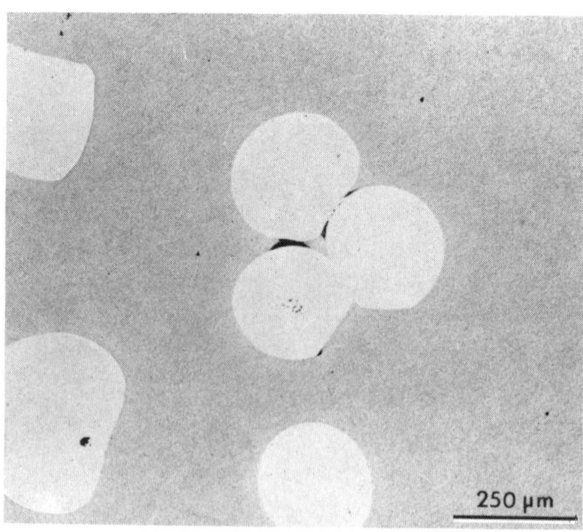

Fig. 2. Microstructure of extrusion 1 as-extruded longitudinal section. Note pores at Nb-Cu interface.

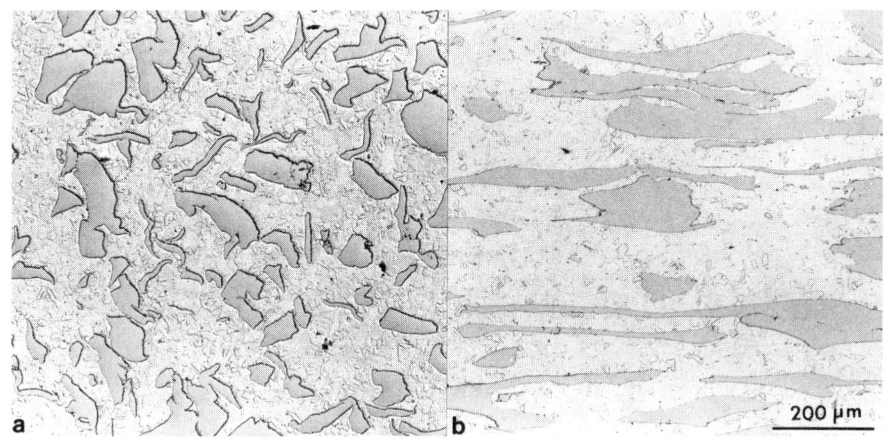

Fig. 3. Microstructure of extrusion 2 as-extruded (a) cross section, (b) longitudinal section.

Figure 5 shows qualitatively and Figure 6 shows quantitatively how the microstructure is developing during the deformation process. Filaments get finer and closer together as deformation increases. Figure 7 is a plot of σ_{UTS} vs. $\bar{\lambda}$, filament spacing, and $\bar{t}$, filament thickness, for both extrusion 2 and a cast composite [4]. The slopes of the lines are each -1/2 indicating that in both materials the Hall-Petch relationship is being obeyed. Some points at large $\bar{\lambda}$'s in extrusion 2 do not fall on the lines; this is due to the added strengthening of the refined grain size. If that incremental strength was subtracted out, the points would fall on the lines. Beyond this region of the plot, a given $\bar{\lambda}$ (or $\bar{t}$) yields the same strength for both the cast and P/M composite.

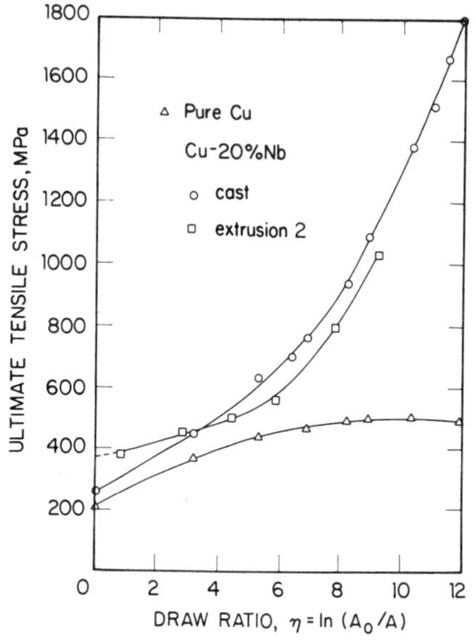

Fig. 4. Effect of deformation on the ultimate tensile strength.

(a)

(b)

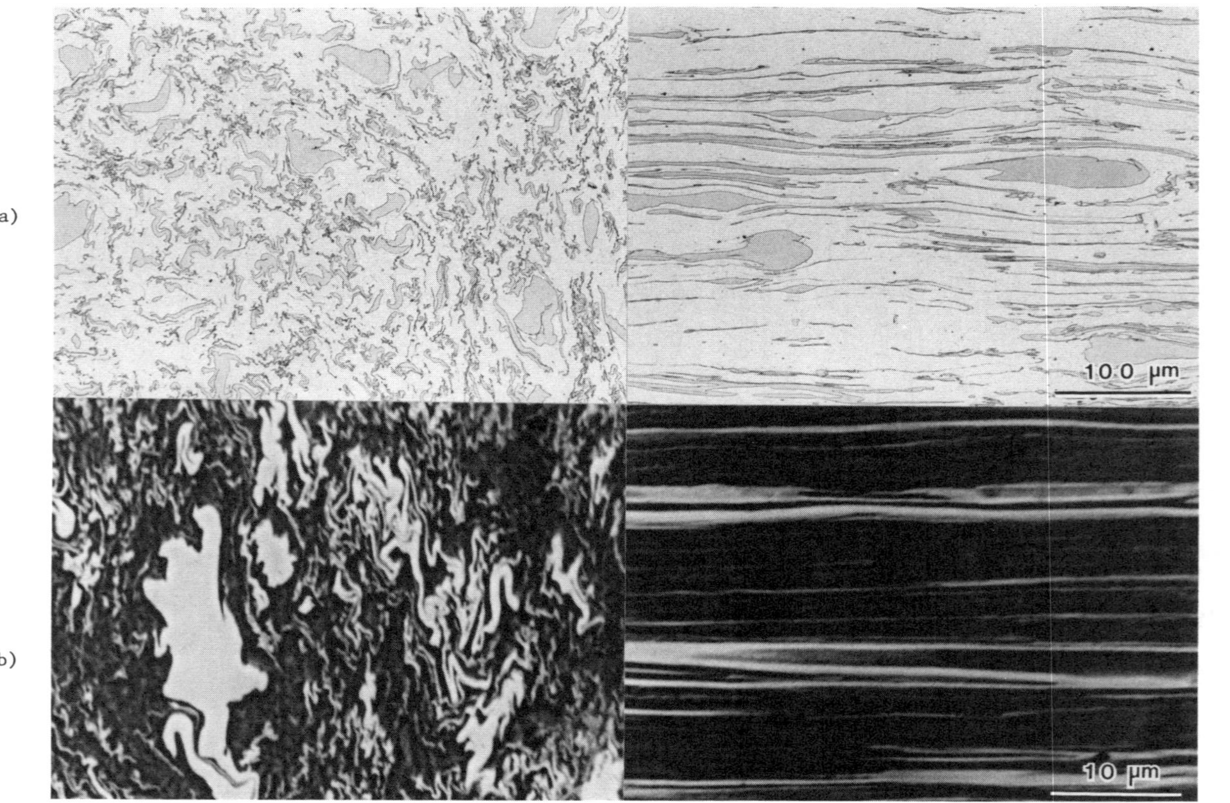

(a)

(b)

Fig. 5. Microstructural evolution of extrusion 2 cross sections (left), longitudinal sections (right). (a) η = 4.5 material, optical; (b) η = 8.0 material, SEM. Nb appears light while Cu is dark.

Table 2. Oxygen Content of Initial Powders and Extrusions

	Oxygen Content (ppm, wt)	(at.%)
Extrusion 1		
Cu	190	0.075
Nb	120	0.070
Composite*	240	---
Extrusion 2		
Cu	790	0.313
Nb	350	0.202
Composite*	980	---

*The oxygen level was measured in the as-extruded material, which contained both Cu and Nb, thus no atomic percent is reported.

Results from a TEM analysis of a cast and wire drawn Cu-20 vol% Nb composite [4,16] show that the Cu undergoes a deformation-recovery-recrystallization cycle which allows for further Nb deformation and refinement. Large populations of dislocations were not observed, but rather a cellular microstructure exists at lower reductions ($\eta < 5$). As intermediate deformations are reached ($5 < \eta < 9$) cells and recrystallized grains were observed while at high η's the microstructure consisted mainly of recrystallized grains. It is believed that the same type of cyclic behavior occurs in the hot extruded P/M composite because in both cases the same microstructural property, $\bar{\lambda}$ of Nb, appears to be controlling the strength. For a given $\bar{\lambda}$ the P/M composite possesses the same strength as the cast one indicating further that similar substructures exist in both materials. Thus, in each case the Nb is acting as planar barriers to dislocation motion and the Cu, by undergoing a cold cycle of deformation-recovery-recrystallization allows for its continual deformation without need for intermediate heat treatments.

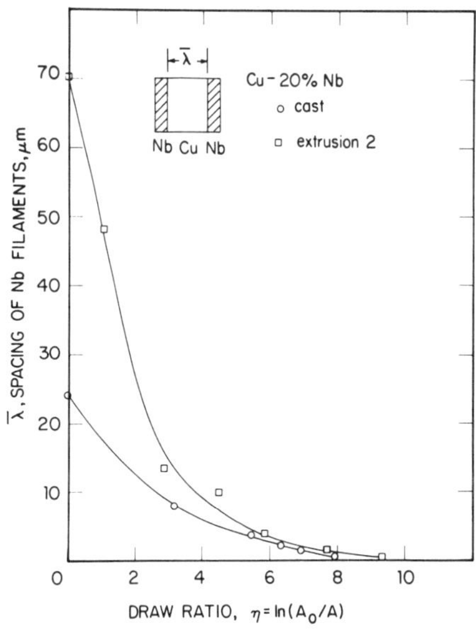

Fig. 6. Illustration of the refinement in filament spacing for cast and extruded composites.

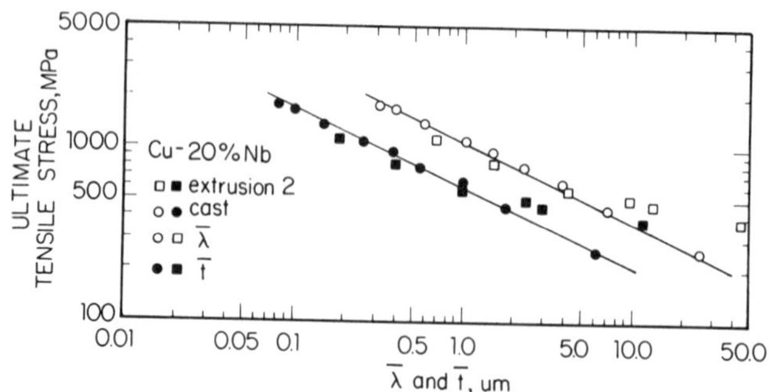

Fig. 7. Log-log plot of strength as a function of $\bar{\lambda}$ and $\bar{t}$.

Conclusions

1. Both the cast and P/M processed Cu-Nb composites show dramatic increases in strength with increased deformation.

2. Strengthening in both cast and P/M composites follows a Hall-Petch relationship which indicates that the strength arises from Nb filaments acting as planar barriers to dislocation motion.

3. Microstructures of both the cast and P/M composites appear to evolve in a like manner with deformation because both yield similar mechanical properties.

4. It has been demonstrated that the P/M route is a feasible method for the fabrication of high strength in situ composites.

Acknowledgments

The experimental assistance of Mr. L. K. Reed and Mr. E. D. Gibson and the metallographic contributions of Mr. H. H. Baker are gratefully acknowledged. The work was performed at Ames Laboratory operated for the U. S. Department of Energy by Iowa State University under contract No. W-7405-ENG-82.

References

1. J. Bevk, J. P. Harbison and J. L. Bell, J. Appl. Phys., 49(12)(1978), 6031-6038.
2. F. D. Lemkey, H. E. Cline and M. McLean, eds., In Situ Composites IV, (New York, NY: Elsevier Science Publishing Co., 1982), 121-133, J. Bevk, et al.
3. W. F. Hosford, Jr., Trans. Met. Soc. AIME, 230(1964), 12-15.
4. W. A. Spitzig, A. R. Pelton and F. C. Laabs, accepted for publication in Acta Met (1987).
5. J. D. Verhoeven, F. A. Schmidt, E. D. Gibson and W. A. Spitzig, J. of Metals, 38(9)(1986), 20-24.
6. R. Borman, H. C. Freyhardt and H. Bergmann, Appl. Phys. Lett., 35(12) (1979), 944-946.
7. L. Schultz and R. Bormann, J. Appl. Phys., 50(1)(1979), 418-424.
8. M. Suenaga and A. Clark, eds., Filamentary A15 Superconductors (New York, NY: Plenum Press, 1980), 289-298, H. C. Freyhaudt, R. Bormann and K. Mroviec.

9. Reed and Clark, eds., Advances in Cryogenic Engineering Materials, vol. 28 (New York, NY: Plenum Press, 1982), 41-47, S. Foner.
10. J. Otubo, S. Pourrahimi, H. Zhang, C. L. H. Thieme and S. Foner, Appl. Phys. Lett., 42(5)(1983), 469-471.
11. R. Flükiger, R. Akihama, S. Foner, E. J. McNiff, Jr., and B. B. Schwartz, Appl. Phys. Lett., 35(10)(1979) 810-812.
12. K. Tachikaua and A. Clark, eds., ICEC-ICMC Conference Japan 1982 (England: Butterworths, 1982), 3119-3122, S. Pourrahimi, et al.
13. R. Flükiger, S. Foner, E. J. McNiff, Jr. and B. B. Schwartz, Appl. Phys. Lett., 34(11)(1979) 763-766.
14. C. L. Trybus, J. D. Verhoeven, F. A. Schmidt and W. A. Spitzig, in preparation.
15. E. E. Underwood, Quantitative Stereology, (Reading, MA: Addison Wesley, 1970), 48-104.
16. A. R. Pelton, F. C. Laabs, W. A. Spitzig and C. C. Cheng, Ultramicroscopy, 22 (1987) 252-266.

TENSILE STRENGTH - MICROSTRUCTURE CORRELATIONS

IN SILICON CARBIDE/ALUMINUM COMPOSITES

C. R. Harris and F. E. Wawner

Department of Materials Science
University of Virginia
Charlottesville, VA 22901

Abstract

An investigation of thermal and mechanical processing effects on Al-20 v/o SiC whisker composites has been made. Differential scanning calorimetry was used to identify temperatures where precipitation or other reactions occur. Tensile specimens were heat treated at temperatures where reactions were found. Room temperature strength properties were found to vary in a non-linear and not necessarily decreasing manner with increasing heat treatment temperatures. Electron microscopy was used to identify phases present after heat treatments. Microstructural characteristics were then correlated with the observed variations in tensile properties.

Processing and Properties for Powder Metallurgy Composites
Edited by P. Kumar, K. Vedula and A. Ritter
The Metallurgical Society, 1988

Introduction

Discontinuous silicon carbide (SiC) whiskers are added to aluminum matrices to improve the strength properties of the material. Higher modulus, yield strength, and ultimate tensile strength can be obtained (1). There are several theories offering explanations for the physical basis of this phenomenon. One theory is based on the generation of dislocations due to the difference in thermal expansion coefficients between the Al matrix and the SiC whiskers (2). When the composite is cooled from high temperatures during fabrication, residual stresses are introduced and dislocations are generated, thus providing the major strengthening mechanism. The whiskers can also act as dislocation traps, inhibiting slip during deformation.

A second theory is that of composite strengthening(3). The modulus of the SiC is much higher than that of the aluminum matrix. Stresses applied to the composite are transferred through the matrix to the whiskers, which can support a much greater load. This theory is generally accepted for long, continuous fibers, but there is a question as to whether or not relatively short whiskers (5 to 10 μm) with an average aspect ratio of 3 or 4 are of sufficient dimensions to support a load.

The authors of this paper support the latter theory as the major strengthening mechanism in Al/SiC composites, and in this paper present evidence supporting this contention.

Experimental Procedure

Extruded composites of 1100 and also 2124 Al matrices with 20 v/o SiC whiskers were obtained from ARCO Metals in Greer, S.C. Aspect ratios of the whiskers were measured metallographically. Some of the materials were also hot-rolled to 55% reduction in thickness.

Tensile tests were performed using an Instron machine. At least five tests were repeated at each test condition. The extrusion processing of these materials tend to align the whiskers along the extrusion direction. When the tensile stresses were applied in this direction, the test is referred to as longitudinal. Transverse tensile tests were also performed with the tensile axis oriented 90 degrees to the extrusion direction.

Fracture surfaces were examined with a JEOL model JSM-35 scanning electron microscope. Sections of fractured specimens were also prepared for transmission electron microscopy (TEM). Disks were cut and mechanically polished before final preparation using ion-milling techniques. A Philips EM400 microscope, operated at 120 KV, was used to examine specimens.

Results and Discussion

Uneven distribution of the reinforcement can result in detrimental tensile properties. The fracture surface in Figure 1 shows areas of matrix ductility in whisker-free areas. Similarly, decreased tensile strength results from areas of high whisker density. In Figure 2 regions of unwetted whiskers are shown. These clumps of unwetted reinforcement essentially act as large voids in the material.

An experiment was performed in an attempt to differentiate between the tensile strength contribution of whiskers and the contribution of matrix precipitates. Naturally aged 2124-20 v/o SiC was tensile tested, noting the yield and ultimate tensile strengths. For comparison, the same material was solutionized, quenched, and then immediately tested. In the second case, only the SiC whiskers were present to act as the major strengthening mechanism. Table I lists the results of these tests. When matrix

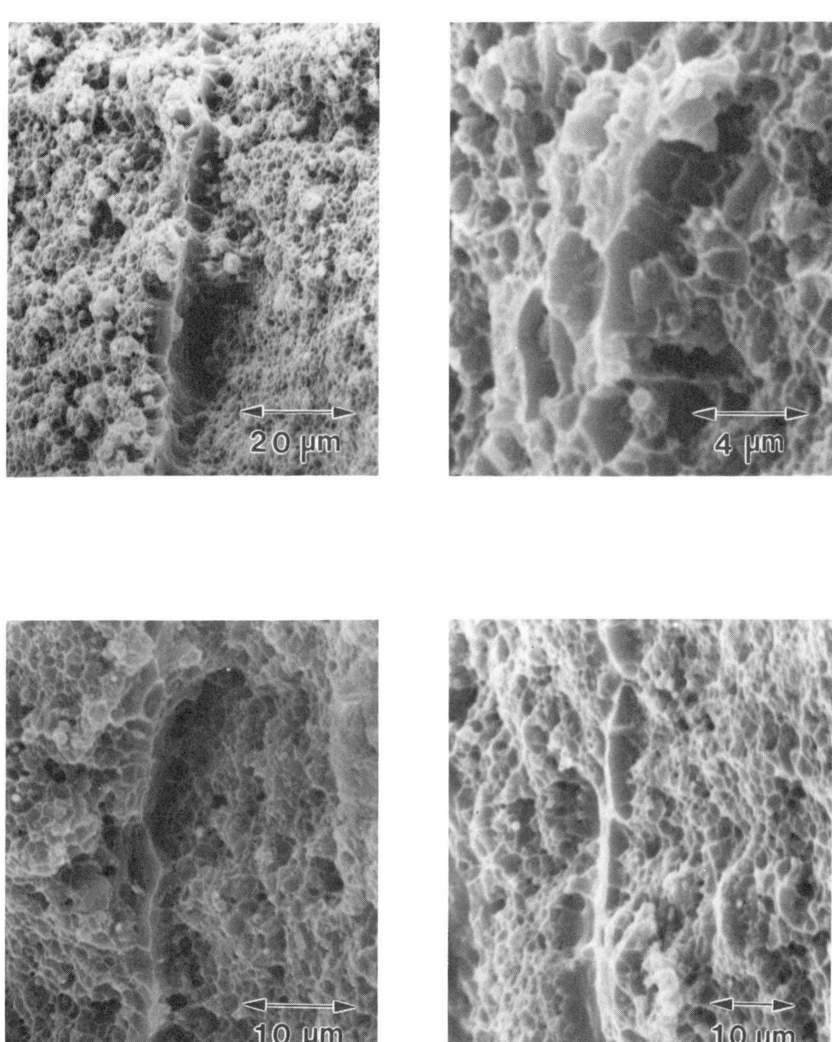

Figure 1. Areas devoid of whiskers showing matrix ductility on composite fracture surfaces.

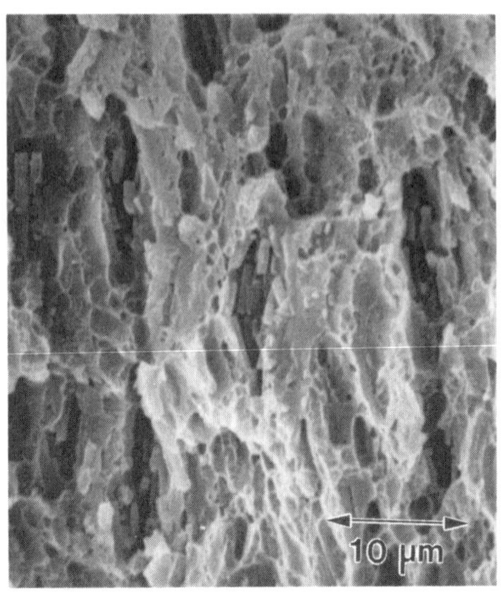

Figure 2. Clusters of unwetted whiskers.

Table I. Comparison of strengths in 2124 composites, with and without matrix precipitates present.

Material	Ultimate Tensile Strength (MPa)	Ultimate Tensile Strength (Ksi)	Yield Strength (MPa)	Yield Strength (Ksi)
2124-T4	717	104	392	56.8
2124: 510°C 1 hr., CWQ (not annealed)	668	97.0	309	44.8

precipitates are not present in the composite material, the yield strength decreases 20% while the ultimate strength shows only a 6% decrease. This indicates that the SiC whiskers make the major contribution to the ultimate strength of the composite, while at lower stresses, precipitation hardening makes a significant contribution to the yield strength of the material.

Measurements of transverse vs. longitudinal strength for several composite systems reveal that the ultimate strength transverse to the whisker alignment axis (extrusion direction) is generally 25% less than that in the longitudinal direction. This result is expected from composite theory since the effective aspect ratio in the transverse direction is essentially unity, while the most probable aspect ratio in the longitudinal direction was determined metallographically to be three. Nardone and Prewo (4) have proposed that significant strengthening can be achieved with whiskers of these dimensions.

To further investigate the load carrying role of the SiC whiskers, sections of fractured tensile specimens were examined using TEM. Figure 3 shows whiskers which have fractured under the applied load. The tensile axis orientation is in the general direction of the whisker alignment direction. Areas devoid of matrix material can be seen between the corresponding halves of the fractured whiskers. Also a high dislocation density is evident in the matrix at the corners of the whisker segments where extensive matrix deformation has occurred. If the whiskers had broken during fabrication, the region between halves should have been infiltrated with matrix material. Also, examination of undeformed material revealed no similar whisker damage or cavitation between whiskers, although on a much larger scale some porosity is present.

The most probable explanation for the observed whisker fracture is that the tensile load has been transferred from the matrix to the whiskers through a strong interface. Fracture usually occurs in the central region of the whiskers where maximum stress can be developed within the fibers. Any defects along the whisker length can, of course, shift the fracture initiation site away from the central region, but this was not generally observed. Also, no interfacial damage was obvious elsewhere along the whisker lengths. This lack of interface cracking or cavitation is indicative of strong bonding, which is assumed to be necessary for a transfer of stress from matrix to reinforcement.

Samples of extruded 1100 Al-20 v/o SiC whisker composites were room temperature tensile tested after one hour exposure to various elevated temperatures, as shown in Figure 4. The curve for the extruded material indicates that ultimate tensile strength is relatively independent of one hour exposures. The other curve in the figure is for the same 1100 composite, but now the material has been hot-rolled to 55% reduction in thickness. Prior to subsequent heat treatment, the ultimate tensile strength of the material has decreased significantly due to the rolling process. After exposure to increasingly higher temperatures, as in Figure 4, the strength is found to increase again back to its original value before rolling.

Reasons for the decrease in strength after rolling are indicated in Figure 5. The fracture surface of rolled material prior to heat treatment shows extensive whisker pullout (Fig. 5a). Fracture surfaces of rolled material after one hour exposure to temperatures above 350°C no longer show this large extent of pullout (Figure 5b). These observations indicate that rolling damages the interface, resulting in pullout, and that subsequent heat treatment relieves the induced damage. Pullout does not occur and the strength is higher in the heated samples.

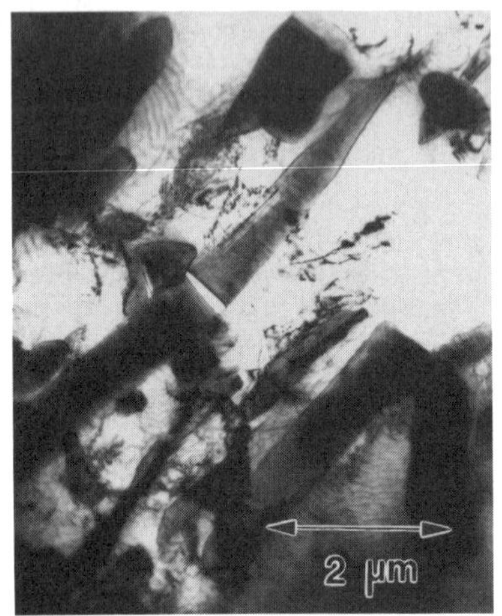

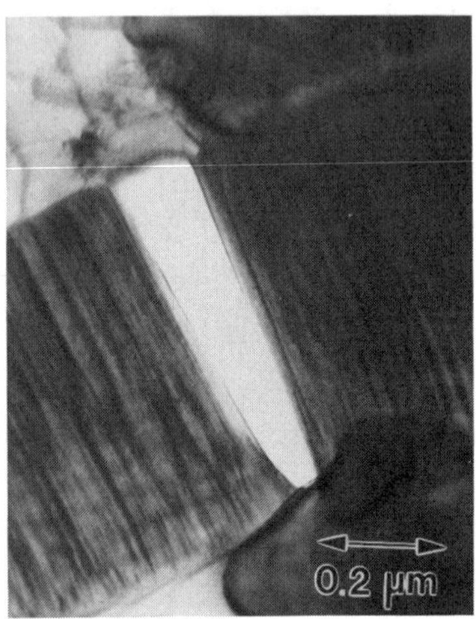

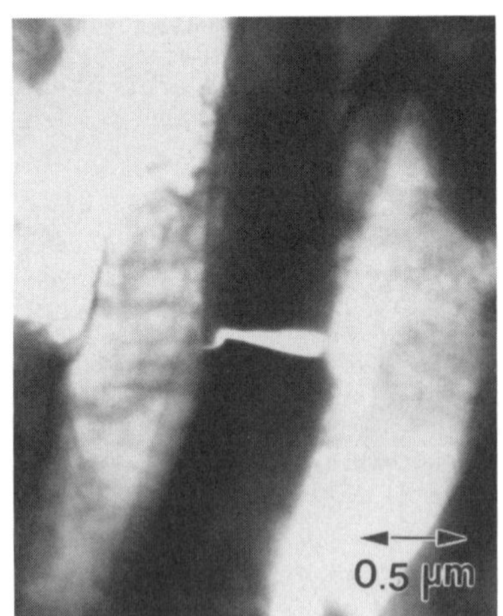

Figure 3. Fractured whiskers in tensile specimens.

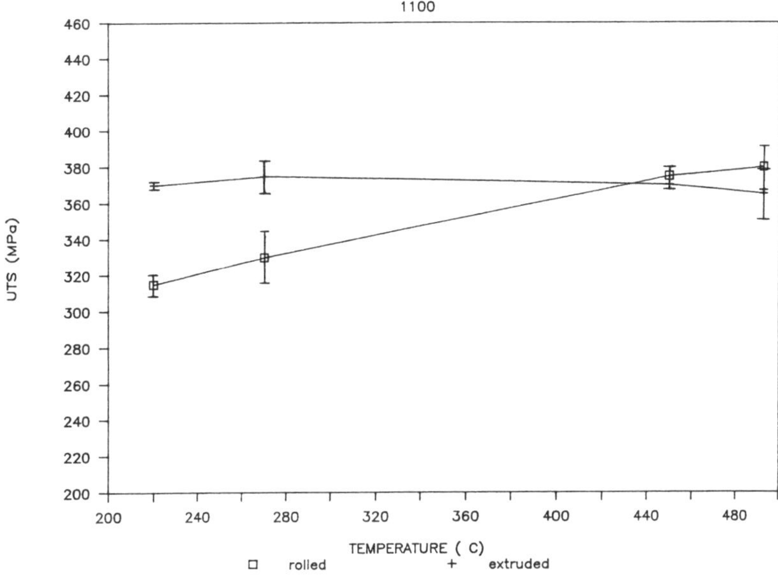

Figure 4. Graph of ultimate tensile strength vs. heat treatment temperature for rolled and extruded 1100 composites.

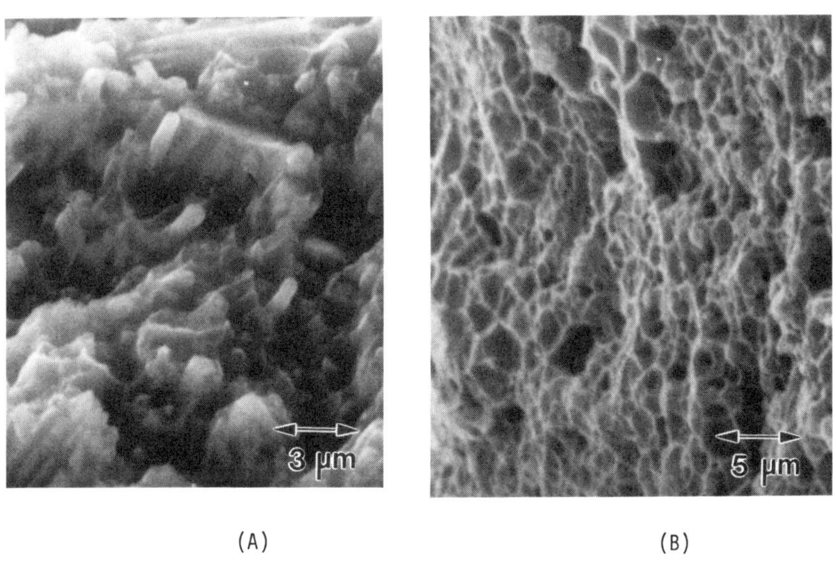

Figure 5. Whisker pull out (A) in as-rolled material. (b) Minimal pullout after heat treating rolled material.

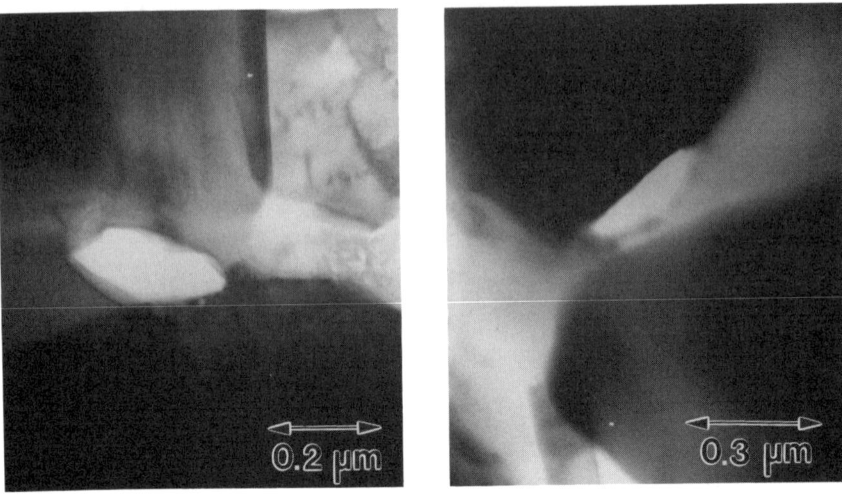

Figure 6. Interfacial damage induced by rolling.

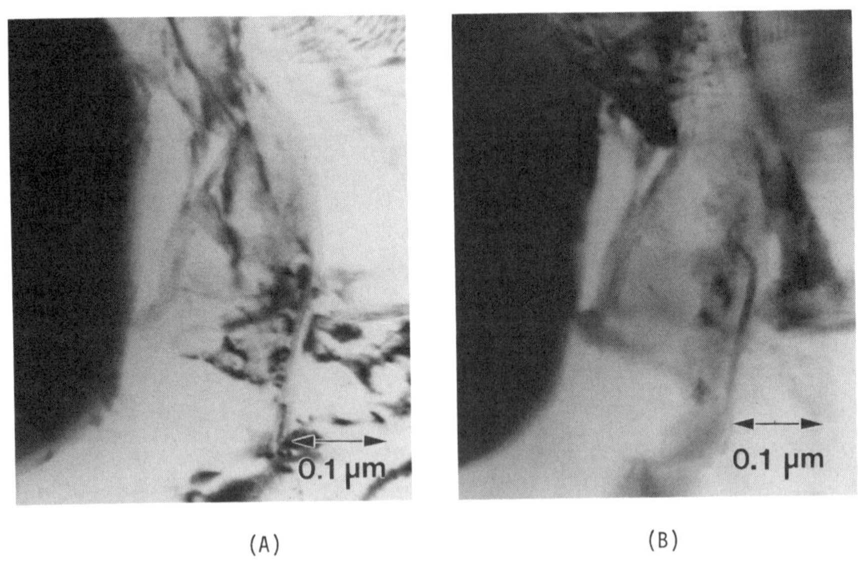

(A) (B)

Figure 7. Series indicating interfacial repair during in-situ TEM heating experiment.

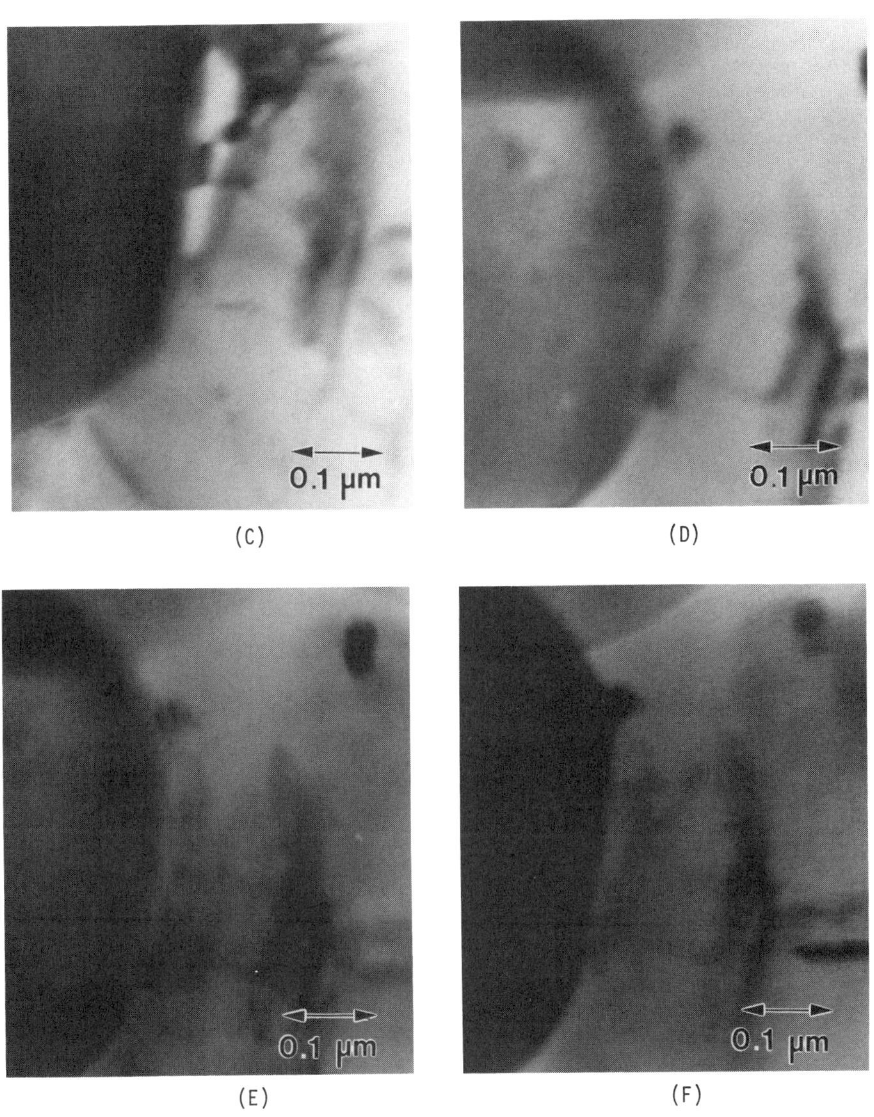

Figure 7 continued. Micrograph series indicating repair of interfacial damage during heat treatment.

TEM observations of the fiber-matrix interface in rolled material show pore formation along whisker cross sections (Fig. 6). Again, similar void formation or cavitation was not observed in undeformed material. In an attempt to gain some insight into what type of process may be occurring at the interface during heating of rolled material, an in-situ heating experiment was done in the TEM. A sample of rolled material, again showing void formation at the interface, is shown in Figure 7a. Figures 7b-e show the same void as time progresses and heat is applied. The figures indicate that the void is being eliminated during heating. Finally, in Figure 7f the void is shown to remain sealed after cooling to room temperature.

Conclusions

Observations of fractured whiskers in tensile composite specimens, along with transverse and longitudinal measurements of ultimate tensile strength indicate that composite strengthening is the major strengthening mechanism in discontinuous SiC/Al composites.

Rolling and heat treatment studies show the importance of the fiber-matrix interface in that damage to the interface can decrease the ultimate strength of materials. However, subsequent heat treatment can relieve induced damage and restore tensile strength to original values.

Acknowledgement

The authors appreciate the support of the Office of Naval Research, Dr. Steven G. Fishman, scientific officer, under grant # N00014-85-K0179.

References

1. A.P. Divecha, S.G. Fishman, and S.D. Karmarker, J.Met., 33 (1981), 12.

2. R.J. Arsenault, R.M. Fisher, Scripta Metall., 17 (1983), p. 67.

3. G.E. Dieter, "Mechanical Metallurgy," 2nd ed., p. 228, McGraw-Hill, 1976.

4. V.C. Nardone and K.M. Prewo, Scripta Metall., Vol. 20, pp. 127-132, 1986.

MICROSTRUCTURAL EFFECTS ON THE FRACTURE MICROMECHANISMS

IN 7XXX Al P/M-SiC PARTICULATE METAL MATRIX COMPOSITES

J. J. Lewandowski and C. Liu
Department of Metallurgy and Materials Science
Case Western Reserve University
Cleveland, OH 44106

W. H. Hunt, Jr.
Alcoa Laboratories
Alcoa Center, PA 15069

Abstract

This paper details a study of the microstructure-property relationships in 7XXX Al P/M matrix-SiC particulate reinforced composites. The focus of this work is on microstructural variables such as the SiC:Al particle size ratio and matrix temper, and their effects on macroscopic fracture behavior through measures such as tensile ductility and fracture toughness. Quantitative fractography and metallography using TEM, SEM, and optical techniques were used to investigate the details of the fracture micromechanisms. It is shown that fracture tends to initiate in clustered regions, regardless of matrix temper, and seeks the larger particles in the distribution during crack propagation. A tendency to shift the fracture mode from particle cracking to decohesion was observed as the matrix temper was varied from underaged to overaged in smooth tensile specimens, while TEM revealed precipitation at the particle/matrix interface in the overaged condition. This change in tensile fracture morphology accompanies a substantial decrease in toughness for the overaged temper, unlike the case for the unreinforced matrix. Possible explanations for the effects of particle clustering and matrix temper are discussed.

Processing and Properties for Powder Metallurgy Composites
Edited by P. Kumar, K. Vedula and A. Ritter
The Metallurgical Society, 1988

Introduction

Metal matrix composite (MMC) materials are currently being explored for use in a range of structural and nonstructural applications. To a great extent, the ability to tailor both their physical and mechanical properties to the needs of the application has stimulated this interest. Discontinuously reinforced MMC's hold the additional advantage of more isotropic properties than continuous fiber materials and may be fabricated using existing metalworking processes.

The mechanical behavior of unreinforced 7XXX aluminum alloys has been extensively studied throughout the years. Figure 1 schematically illustrates the effect of matrix temper on the strength and fracture toughness behavior of a typical unreinforced 7XXX alloy such as 7075 (1). Increases in aging time for the underaged condition initially result in increased strength, accompanied by a decrease in fracture toughness. (From a practical point of view, underaged tempers (designated W) are not stable at room temperature in 7XXX alloys, making it necessary to consider the peak aged (T6) and overaged (T7) tempers only for commercial applications.) The minimum in toughness occurs at the highest strength, peak aged condition. Continued aging, i.e., overaging, results in a loss in strength due to coarsening of the precipitate phases. Although much of the fracture toughness is recovered in the unreinforced overaged materials, the strength-toughness combination is lower than in the underaged condition of comparable strength. A variety of explanations have been proposed for this behavior (1-3). In some aluminum alloys, it has been demonstrated that a fracture mode change from dimpled transgranular to dimpled intergranular on going from the underaged to overaged temper is primarily responsible for the observed drop in fracture toughness (1,2). However, in cases where no change in fracture mode is obtained, it has been proposed (3) that crack tip bifurcation may occur in underaged alloys, thereby decreasing the Mode I component of stress (i.e., driving force) at the crack tip. In contrast,

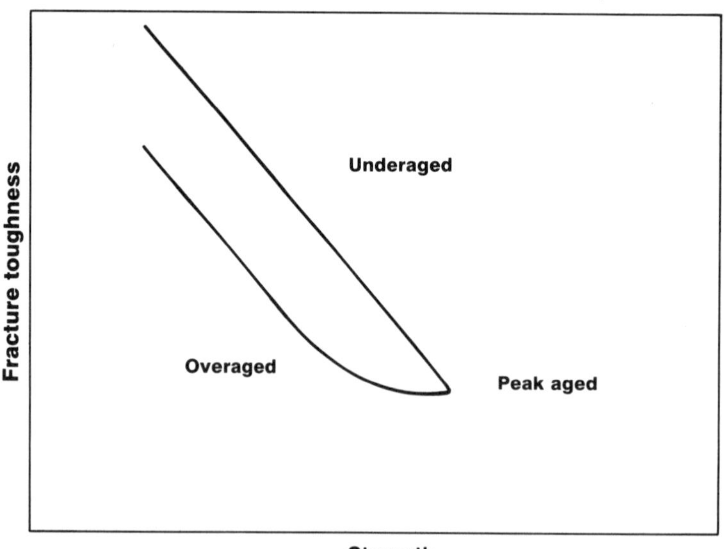

Figure 1 - Typical toughness-strength behavior as a function of aging for unreinforced 7XXX Al alloys.

the overaged tempers propagate in an essentially Mode I manner. Still others have focused on the influence of the coarser precipitate phases in the overaged structure and their effects on the ductile fracture process by reducing void initiation and/or growth strains.

Due to the recent interest in aluminum matrix MMC's, and as a complement to the work conducted on unreinforced aluminum alloys, the goal of the present work was to investigate the fracture micromechanisms responsible for the observed strength-toughness relationships for MMC's based on a 7XXX matrix. While substantial engineering data have been generated on some of the MMC systems, the understanding of the relationships between the composite microstructures and mechanical properties is somewhat less developed. The focus in the present and continuing work is on developing this structure-property knowledge, initially in aluminum matrix-silicon carbide particulate MMC systems. Previous work (4) demonstrated that particulate size and volume fraction, as well as matrix temper, had significant effects on the strength-fracture toughness relationships in 7XXX P/M Al matrix-SiC particulate extrusions, although the details of the fracture processes were not investigated. The present work represents a continuation of those preliminary studies, with the goal of quantifying the effects of these variables on the mechanical behavior of this class of composites.

Experimental Procedure

Fabrication

Figure 2 schematically shows the processing route utilized for the P/M based composites investigated in this study. The 7XXX matrix alloy composition used in this work, designated Alcoa MB78, contains 7% Zn, 2% Mg, 2% Cu, and 0.14% Zr, balance Al, and was reinforced with 20% by volume of either F-600 grade (average size ≈16 μm) or F-1000 grade (average size ≈5 μm) SiC produced by Norton Co. The alloy powder was produced by an air atomization process, with subsequent screening to produce 90% -325 mesh. An initial surface conditioning treatment is followed by a proprietary dry blending of the aluminum powder and SiC particulates to produce a material reinforced with SiC particulate that is well dispersed. It has been shown that the relative sizes of the starting powders influences the randomness of the final particulate distribution (5). In the present work, the controlling parameter is the average SiC particulate to aluminum powder particle size ratio (SiC:Al PSR). For the size distributions of particles of the starting SiC particulate, shown in Figure 3, SiC:Al PSR's of 0.3:1 and 0.7:1 for the F-1000 and F-600 SiC materials, respectively, are calculated. Increasing the difference in the relative particle sizes (i.e., lower SiC:Al PSR) results in greater segregation of the two constituents, which is schematically illustrated in Figure 3. One of the goals of the present work was to evaluate the effect of two different SiC:Al PSR levels on the microstructure and resultant mechanical properties.

After blending, the powders were cold compacted to approximately 75%-80% of theoretical density and subsequently degassed and vacuum hot pressed to near (i.e., 99%) theoretical density. All of the compaction processes were carried out at temperatures below the solidus temperature of the matrix. This is done to insure the retention of the rapidly solidified microstructure and prevent the formation of coarse intermetallic phases often observed in MMC materials produced using supersolidus hot pressing temperatures (6). The billets were subsequently fabricated by direct extrusion to a rectangular bar of cross section 25.4 mm x 76.2 mm at an extrusion ratio of 21:1.

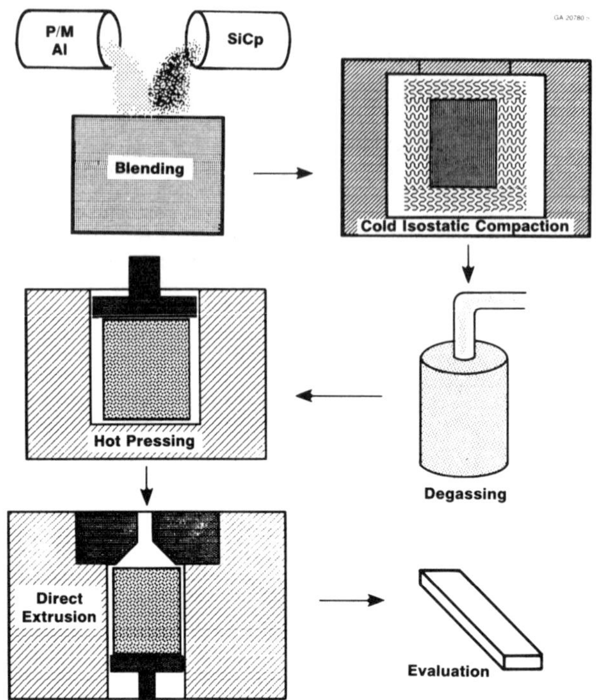

Figure 2 - Processing route for P/M Al-SiC particulate metal matrix composites.

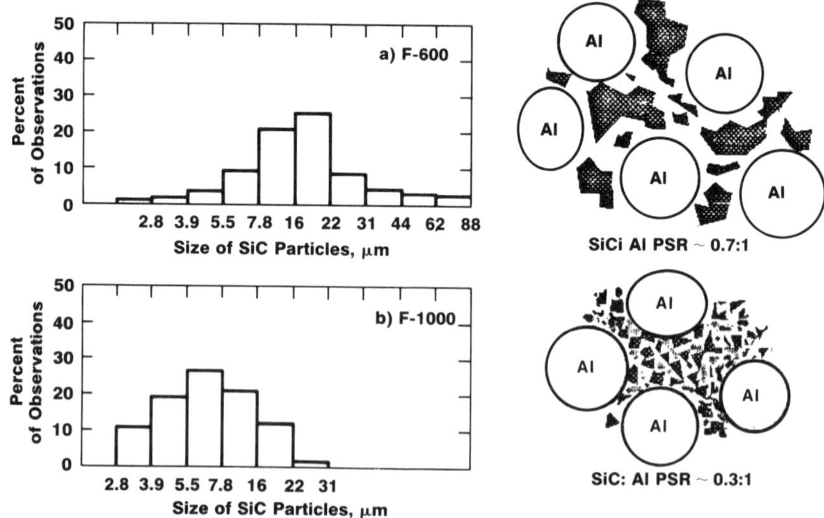

Figure 3 - Size distributions of SiC particulates prior to blending with accompanying illustration of the effect of SiC:Al PSR on resultant distribution in the blend, a) F-600 grade, and b) F-1000 grade.

Heat Treatment

The as-extruded materials were subsequently solution heat treated, cold water quenched, and aged in order to systematically vary the matrix microstructure. The heat treatments are listed in Table I for the two SiC:Al PSR composite variables as well as the unreinforced control material. In particular, the aging practices were selected to match the strengths of the materials in the underaged and overaged tempers in order to better study the effect of microstructure on the fracture behavior. Future references to matrix temper in this paper will use W1 and W2 to designate the underaged tempers, T6 the peak aged temper, and T7X1, T7X2, and T7X3 the overaged tempers, as defined in Table I.

Table I. Aging Practices for MB78 Matrix Composite Materials[1]

Temper Designation	Material		
	MB78 Control	MB78 + 20% F-600 SiCp	MB78 + 20% F-1000 SiCp
W1	4 h @ 25°C	4 h @ 25°C	4 h @ 25°C
W2	96 h @ 25°C	168 h @ 25°C	168 h @ 25°C
T6	24 h @ 121°C	24 h @ 121°C	24 h @ 121°C
T7X1	T6 + 6 h @ 177°C	T6 + 12 h @ 177°C	T6 + 12 h @ 177°C
T7X2	T6 + 24 h @ 177°C	T6 + 48 h @ 177°C	T6 + 48 h @ 177°C
T7X3	T6 + 48 h @ 177°C	T6 + 72 h @ 177°C	T6 + 72 h @ 177°C

1 All materials solution heat treated for 4 h @ 530°C and quenched in 25°C water prior to aging.

Microstructural Evaluation

Quantitative metallographic techniques were utilized to characterize the material with respect to average particle size, particle size distribution, and particle clustering. Actual particle size distributions were determined on metallographic sections using a Videoplan Image Analysis Microscope. Features obtained at 1000X-2000X were analyzed to generate data such as average particle size, aspect ratio, and size distribution. In addition to quantitative metallography, the microstructures were analyzed using TEM on a JEOL 200CX. Thin foils were prepared by ion milling in an ion miller equipped with a cold stage.

Clustering of particles was determined quantitatively by using the combination of data acquisition with an automated scanning electron microscope followed by data analysis by the Dirichlet tessellation technique (7). Metallographically mounted specimens were imaged in the backscattered electron mode in an ISI 100 SEM. Data for ≈1000 particles, consisting of the location of the centroid of the particle in the field, particle area, particle width, particle length, and particle orientation relative to the horizontal, were acquired by the LeMont system interfaced with the SEM. This data was then transferred to a VAX computer and processed by a series of programs designed to perform the Dirichlet tessellation analysis along with associated functions such as statistical analysis and graphics.

The Dirichlet tessellation analysis essentially involves the division of the microstructure into Dirichlet cells, with each cell containing a single particle and that area of matrix which is closer to the particular particle than any other. An example of the reconstruction of a microstructure with the Dirichlet cells shown is presented in Figure 4. The current version of the program approximates the particles as circles and creates the tessellated structures using the particle centroid data. This results in some inaccuracy and particle overlap in regions of the structure where particles are closely spaced. Improved versions which incorporate the particle aspect ratio as well as eliminate the particle overlap problem are currently being developed. For the purposes of this work, however, in which the relative degree of particle clustering is sought, the current version is adequate.

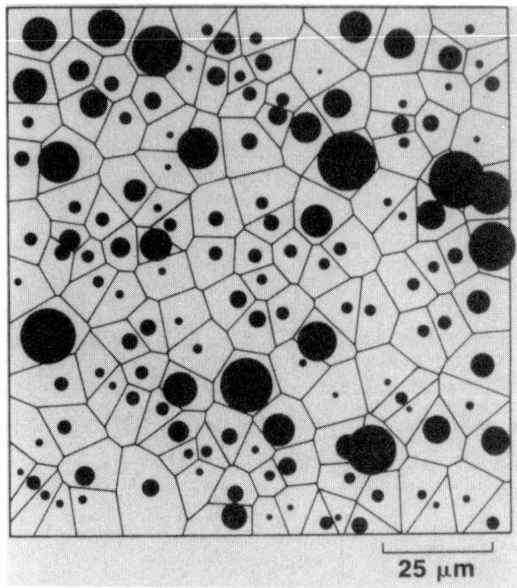

Figure 4 - Computer representation of MMC microstructure after Dirichlet tessellation analysis.

The Dirichlet construction provides a number of useful metallographic parameters. Near neighbors are explicitly defined as those sharing a cell boundary with the subject particle, with the nearest neighbor being the near neighbor at the shortest distance. Local area fraction, defined as the ratio of the particle area to the cell area surrounding it, and the distribution of the local area fractions are also provided. These are especially useful for the consideration of particle clustering, as discussed later. In addition, the cell boundaries provide an indication of the extent of plastic zone growth from a particle before impingement with the plastic zone of a neighboring particle, which provides some physical insight into the importance of clustered particles in the deformation and fracture processes in these materials. Finally, Dirichlet tessellations can be constructed in the computer for microstructures with random distributions of particles with the same global volume fraction as well as mean particle size and standard deviation as actual distributions. This allows the microstructural parameters such as those discussed above to be measured on the random, computer-generated microstructure and compared to the actual microstructure in order to quantitatively assess differences.

Mechanical Testing

Tensile testing was performed in the longitudinal orientation on specimens given the various heat treatments outlined in Table I. Cylindrical tensile specimens, with a 3.8 mm diameter reduced section and 15.2 mm gage length were tested at strain rates of ≈0.05/min. Previous internal tests at Alcoa Laboratories had demonstrated that there was no effect of specimen size on properties. Therefore, this smaller specimen was selected to make most efficient use of the available material. Tensile strength, yield strength, elongation, and reduction of area were obtained along with work hardening information.

The bulk of the toughness testing in this program employed the short rod fracture toughness test. The short rod test is a useful indicator of toughness which employs less material and a simpler testing methodology (e.g., no precracking required) than the compact tension approach. In the short rod test, 25.4 mm diameter rods oriented in the T-L orientation were machined with chevron notches, as shown in Figure 5. The specimens were stressed in a loading-unloading manner until fracture propagated, and a short rod fracture toughness value determined as the stress intensity at maximum load. Further information on this technique can be found in Reference 8.

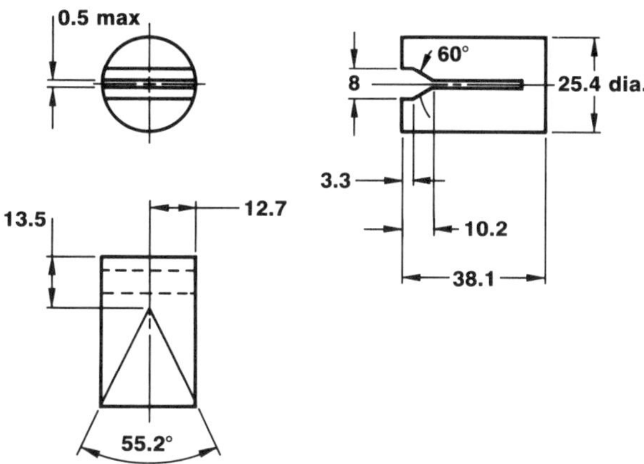

Figure 5 - Specimen design for short rod fracture toughness test. Dimensions are in mm.

Selected materials in the W2, T7X2, and T7X3 tempers were also tested using conventional compact tension specimens according to ASTM E-399, primarily to determine the correlation with the values obtained in the short rod test. In this work, the compact tension and short rod derived values agreed well at toughness levels below ≈25 ksi$\sqrt{\text{in}}$, with the short rod overpredicting the actual toughness by about 10% at higher toughness levels, although this is based on relatively few data.

Quantitative Fractography

Fracture surfaces were analyzed using a JEOL 35 SEM equipped with PGT Microanalysis System. Analyses of the fracture surfaces were performed by matching surface fractography, as well as by quantification of the area fraction and particle size distributions of SiC particulates present on the fracture surface. Observations were made at magnifications of 1000X-5000X. Particular attention was directed toward examining for evidence of SiC particulate cracking vs. decohesion, and the effects of matrix microstructure on the fracture morphology.

Longitudinally sectioned tensile specimens were additionally examined after testing using optical metallography to determine the influence of particulate spatial distribution on fracture.

Results

Microstructure

Figures 6a and 6c show typical microstructures taken parallel to the extrusion direction of the MB78 matrix composites with 20% by volume F-600 grade and F-1000 grade SiC particulates, respectively. The corresponding size distributions of the SiC particles in Figures 6a and 6c are given in Figures 6b and 6d, respectively. Comparison of the particulate size distributions in the fabricated materials in Figure 6 to the starting SiC particulate size distributions prior to blending, shown in Figure 3, illustrates that processing caused particulate fracture. The arrows in Figure 6a illustrate some of these fractured particles. Particulate fracturing occurs to a greater degree in the F-600 SiC reinforced materials as compared to the F-1000. For example, fabrication produces a shift in the average particle size in the F-600 starting particulate case from ≈16 μm to ≈5 μm, and in the F-1000 size from ≈5 μm to ≈4 μm. These micrographs also show that the aspect ratio of the particles is ≈2-4:1, with the lower aspect ratios in the F-1000 SiC composites, along with some preferred orientation along the extrusion direction.

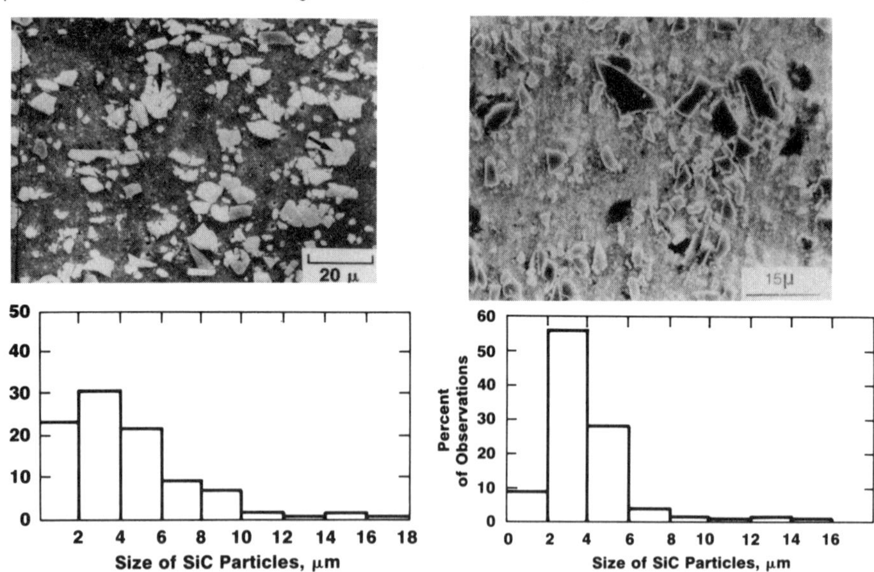

Figure 6 - SEM micrographs and corresponding particulate size distributions of extruded MB78 + 20% by volume SiC particulate composites, a) and b) F-600 grade, c) and d) F-1000 grade. Arrows in a) denote particles fractured during processing.

The quantitative description of clustering of the SiC particulate was determined using the tessellation technique on microstructures such as those shown in Figure 6. Local area fractions were calculated using the procedure described earlier. One method of determining the degree of particle clustering is to compare the distribution of local area fractions observed with that expected for a random distribution. Normalizing the local area fractions by their mean allows comparison of materials with different nominal mean area fractions on the same plot. Figure 7 shows a plot of the cumulative probability of finding a particular normalized local area fraction in the microstructure vs. the normalized local area fraction. The solid line represents the expected values based on a random distribution, while the two dashed lines represent the data for the two MB78 + 20% by volume SiC particulate composites. The displacement of the curves for the composite materials to higher cumulative probabilities at the high end of the normalized local area fraction values indicates greater particle clustering than would be expected in the random case. A comparison of the two composite materials reveals that the SiC:Al PSR = 0.3:1 material exhibits more clustering than the SiC:Al PSR = 0.7:1 material, consistent with the schematic model shown in Figure 3.

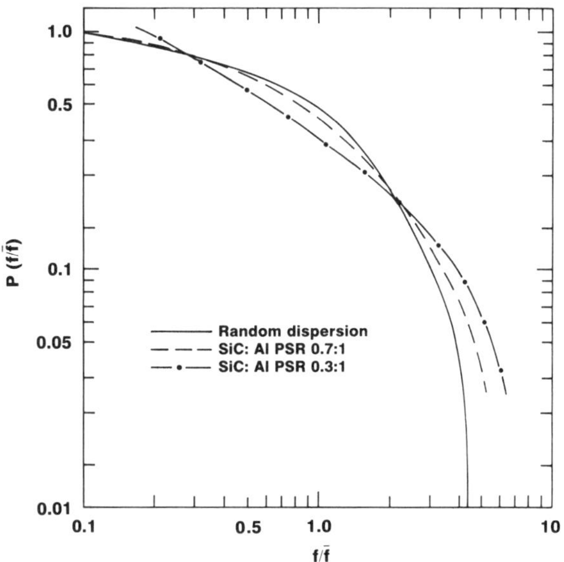

Figure 7 - Cumulative probability plot of normalized local area fractions in MB78 + 20% by volume SiC particulate composites, illustrating clustering tendency.

TEM micrographs of the as-extruded and overaged microstructures are shown in Figures 8 and 9, respectively. In the as-extruded microstructure, Figure 8, the SiC particles themselves were often fractured, but well-bonded to the matrix. The interface region and matrix was heavily dislocated in the as-extruded material, and the interface did not appear to contain precipitates. Defects within the SiC were also observed, as seen in Figure 8b. In contrast, the overaged microstructure exhibited the presence of equilibrium η ($MgZn_2$) phase at the interface, at an average size of 0.1 μm, as shown in Figure 9, while the interface regions in specimens heat treated to underaged tempers were precipitate free. Although it had been suggested in previous work that a high dislocation density would be expected at the SiC particulate/matrix interface in these materials due to the difference in coefficient of thermal expansion between the Al and SiC (9), such high dislocation densities were observed only in the as-extruded material.

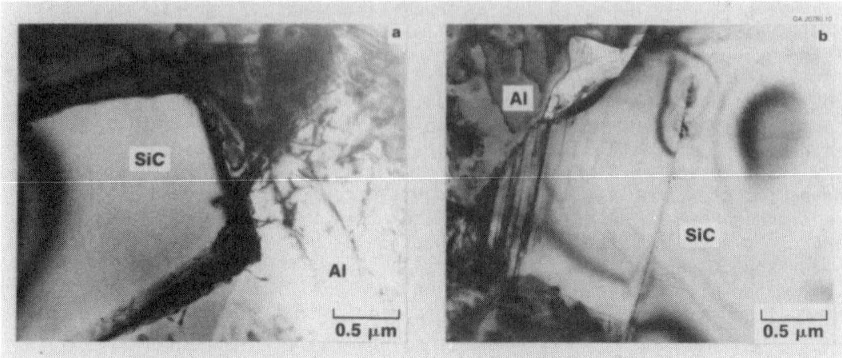

Figure 8 - TEM micrographs of as-extruded MB78 + 20% by volume F-600 grade SiC particulate, showing heavily dislocated matrix in a) and cracked SiC particulate in b).

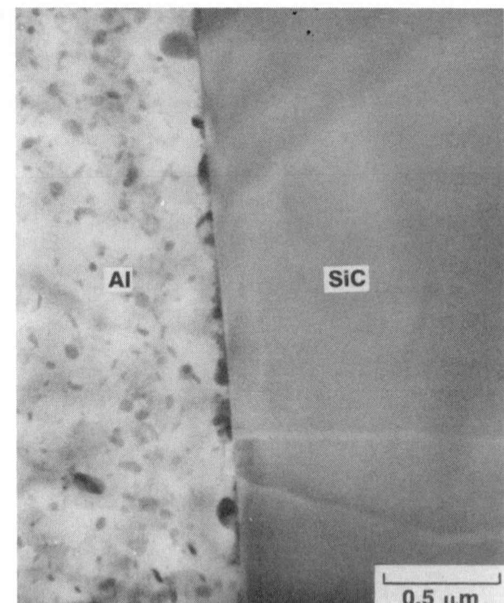

Figure 9 - TEM micrograph showing η ($MgZn_2$) precipitates at the SiC particulate/MB78 matrix boundary in the T7X3 temper.

Mechanical Properties

Table II summarizes the tensile properties obtained in both the unreinforced and composite materials. Compared to the unreinforced MB78, the composite materials exhibit similar strengths, but significantly lower

Table II. Longitudinal Tensile Properties of MB78 Matrix Composite Materials

Temper	MB78 Control				MB78 + 20% F-600 SiCp				MB78 + 20% F-1000 SiCp			
	σ_u (MPa)	σ_y (MPa)	El. in 15.2 mm (%)	RA (%)	σ_u (MPa)	σ_y (MPa)	El. in 15.2 mm (%)	RA (%)	σ_u (MPa)	σ_y (MPa)	El. in 15.2 mm (%)	RA (%)
W1	514	326	23	22	489	325	8	10	498	314	7	10
W2	571	370	19	16	540	389	6	10	542	385	6	8
T6	589	554	17	21	598	543	4	6	583	518	5	5
T7X1	558	516	17	51	561	495	6	8	541	460	6	7
T7X2	429	352	19	52	465	398	5	9	449	345	7	8
T7X3	427	343	18	57	465	387	7	9	429	330	8	9

Table III. Fracture Toughness Properties for MB78 Matrix Composite Materials

Temper	MB78 Control		MB78 + 20% F-600 SiCp		MB78 + 20% F-1000 SiCp	
	$K_{I_{CSR}}$ (1)	K_{I_C} (2)	$K_{I_{CSR}}$ (1)	K_{I_C} (2)	$K_{I_{CSR}}$ (1)	K_{I_C} (2)
W1	62.6(3)	--	27.5	--	23.6	--
W2	56.8(3)	39.7	25.7	23.2	22.5	BDP(4)
T6	31.6	--	15.4	--	14.5	--
T7X1	38.6	--	15.9	--	13.2	--
T7X2	62.7(3)	46.2(3)	15.8	15.5	14.3	13.1
T7X3	66.2(3)	32.2(3)	15.5	15.2	14.4	12.8

(1) Determined by short rod fracture toughness test, T-L orientation.
(2) Determined from compact tension specimens using ASTM E-399 criteria, T-L orientation.
(3) Excessive plasticity -- invalid result.
(4) BDP = broke during precracking.

ductility values, although the latter are excellent for MMC materials at the 20% by volume level (10). Similar values of ductility are observed for the two variables of SiC:Al PSR. Ductility decreases with increased aging up to the T6 temper, then recovers somewhat upon overaging.

Table III shows the toughness data determined from the short rod and compact tension specimens. Note that the high values for the unreinforced matrix material are invalid due to the extensive plasticity in the short rod test. Nevertheless, they give an indication of the behavior of the material, and when considered along with the strength properties, show the expected trend in strength-toughness behavior which had been illustrated in Figure 1. The MMC materials show substantially lower toughness values, although again they are at a high level for this class of material. The important influence of SiC:Al PSR on the strength-toughness relationship is shown in Figure 10, with the larger SiC:Al PSR material showing consistently higher toughness values than the smaller SiC:Al PSR material at a given strength level. Also, in contrast to the unreinforced matrix behavior, the MMC materials do not recover toughness upon overaging. Instead, a relatively low toughness (similar to that for the T6 aged materials) is maintained even in the lower strength overaged alloys (Figure 10).

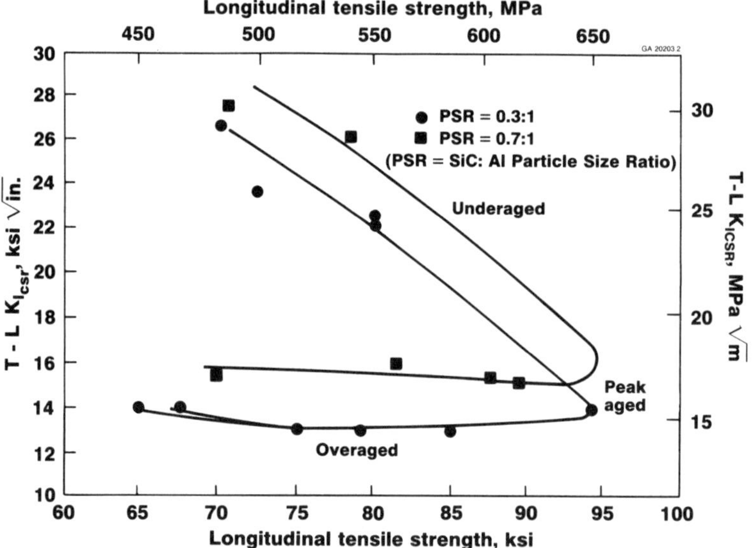

Figure 10 - Toughness-strength relationships for MB78 + 20% by volume SiC particulate composites, showing the importance of SiC:Al PSR and matrix temper.

Fractography and Metallography on Deformed Samples

As a means to qualitatively assess the role of particulate clustering on fracture, metallographically sectioned and polished tensile specimens were studied. The sectioned specimens in Figures 11a and 11b, for the SiC:Al PSR = 0.3:1 and 0.7:1, respectively, show that preferential fracture occurs at clusters of particles with the smaller SiC:Al PSR ratio. In contrast, the fracture path in the more randomly distributed SiC:Al PSR = 0.7:1 material appeared to be more random.

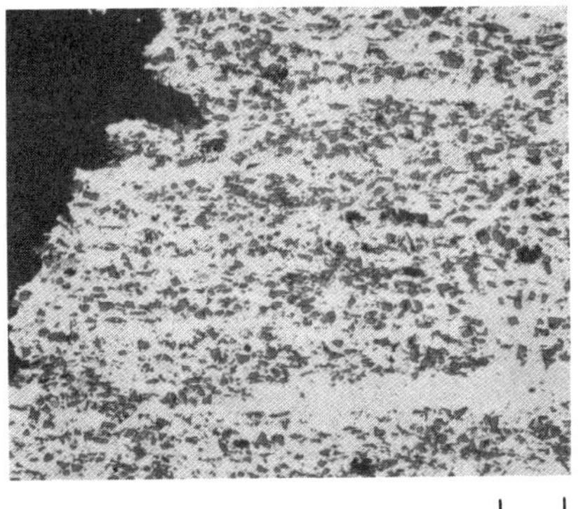

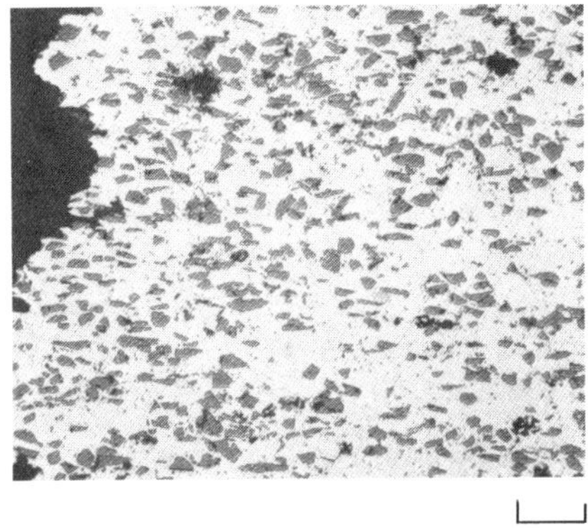

Figure 11 - Longitudinal sections of fractured tensile specimens illustrating the role of SiC:Al PSR on the progress of fracture for a) SiC:Al PSR = 0.3:1 and b) SiC:Al PSR = 0.7:1.

Figures 12 and 13 show typical matching surface fractographs from tensile specimens of the 20% by volume composite with a SiC:Al PSR = 0.7:1 for the W2 and T7X2 conditions, respectively. There was no evidence of inclusion initiated fracture in these specimens, in contrast to other data (6,11). This probably results due to the differences in vacuum hot pressing practice, i.e., subsolidus consolidation in this work vs. supersolidus consolidation in the previous work (5,11).

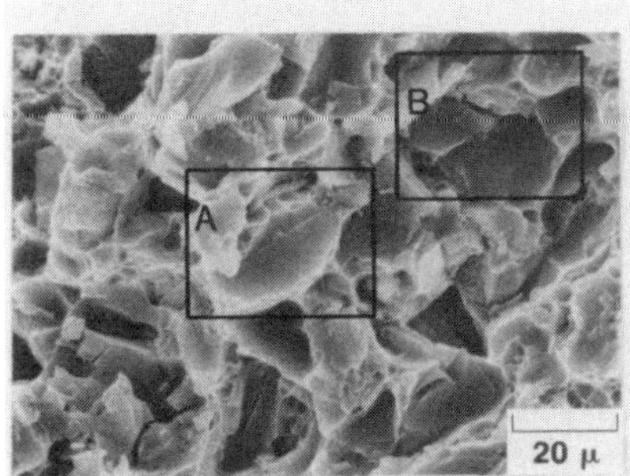

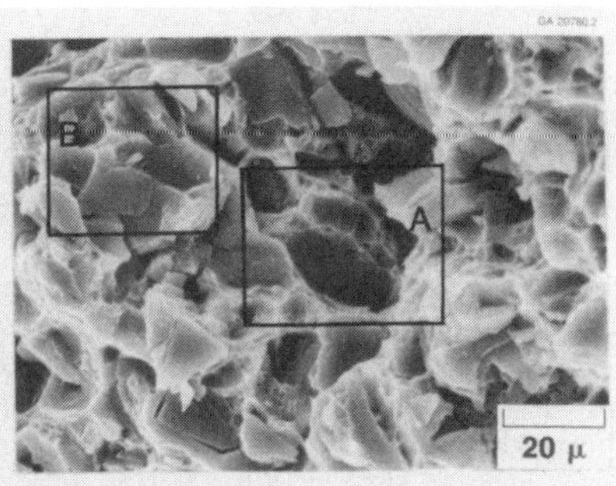

Figure 12 - Matching surface SEM fractographs of tensile fracture in MB78 + 20% by volume F-600 grade SiC particulate in the W2 temper. Particle fracture is seen at A and B.

The tensile specimen in the underaged W2 temper exhibited predominantly cracked SiC particles surrounded by ductile fracture of the matrix. Quantitative fractography on the matching surface fractures revealed that fractured SiC particles accounted for 18% of the fracture surface, with the size distribution of fractured particles as shown in Figure 14a. In contrast, the overaged T7X2 temper material exhibited only

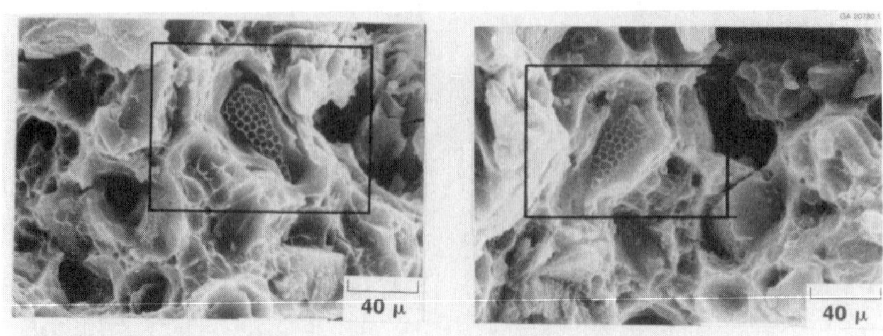

Figure 13 - Matching surface SEM fractographs of tensile fracture in MB78 + 20% by volume F-600 grade SiC particulate in the T7X2 temper. Particle/matrix decohesion is observed in the highlighted area.

11% of the tensile fracture surface occupied by fractured SiC, although the size distribution of fractured particles present on the surface was similar to that of the W2 temper specimen (Figures 14a and 14b). The trend observed for the tensile specimens was that the larger particles cracked preferentially for both aging conditions. This can be more clearly seen by comparing Figures 6 and 14, which represent the particle size distributions before and after tensile testing, respectively. Evidence of decohesion between the SiC particulates and the matrix was obtained in the T7X2 temper specimens as shown by regions of well-defined dimpled fracture at the SiC/matrix interface in Figure 13. Further, TEM examination of foils taken just below the tensile fracture surface of overaged materials revealed void formation associated with the particles at the SiC particulate/matrix interface, Figure 15. These features were not observed in the underaged W2 temper specimens.

Discussion

In discontinuously reinforced metal matrix composites, microvoid coalescence (MVC) is the predominant fracture mode (12). The stages of MVC have been well characterized in a variety of materials (13) and consist of void nucleation, growth, and coalescence stages, the details of which largely determine the fracture properties. It is thus appropriate to extend the current understanding of these ductile fracture processes to investigate the stages controlling toughness in the composites studied presently.

It has been shown that the contribution of void nucleation to the total fracture strain for dispersion hardened alloys undergoing MVC in tensile deformation becomes increasingly significant as the volume fraction of particles increases (14). It would then appear that the details of the void nucleation process should largely control the tensile ductility of aluminum alloys reinforced with SiC particulate at this stage of their development, although void growth and coalescence processes may become more important as the quality of these materials improves.

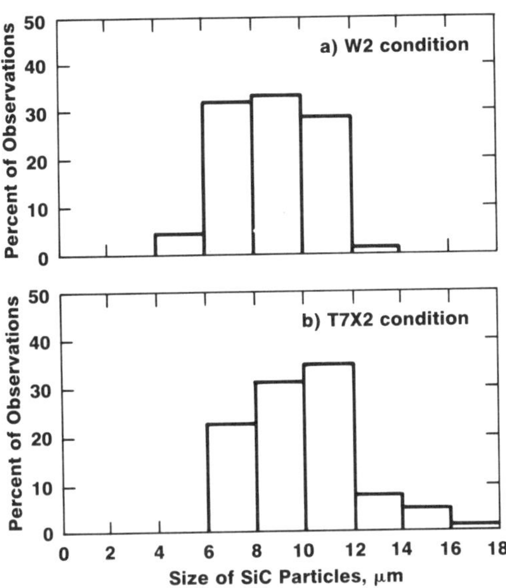

Figure 14 - Particulate size distributions on tensile fracture surfaces in MB78 + 20% by volume F-600 grade SiC particulate composites in the a) W2 and b) T7X2 tempers.

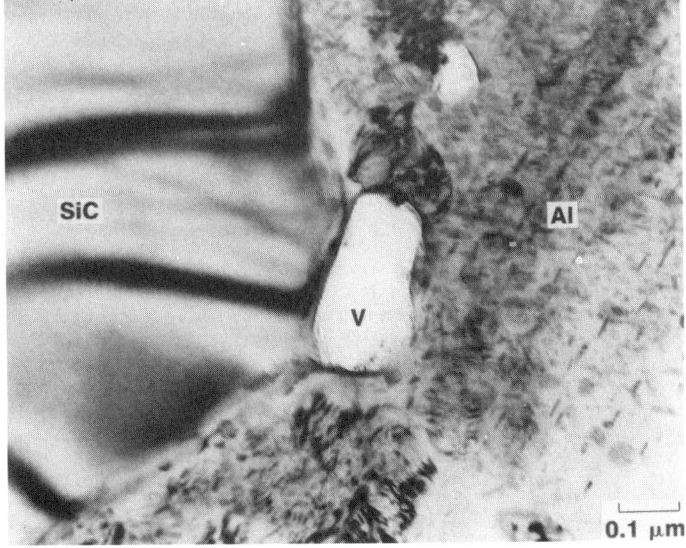

Figure 15 - TEM micrograph taken below the tensile fracture surface showing void formation at the particle/matrix interface (at V) in a MB78 + 20% by volume F-600 grade SiC particulate composite in the T7X2 temper.

From a microstructural viewpoint, the void nucleation process can be influenced by a variety of factors, including the total volume fraction and the local volume fraction of particles; the matrix deformation characteristics; and the particle/matrix interface bond strength. The results obtained in this work indicate that a variety of factors influence the mechanical behavior and fracture properties of SiC reinforced aluminum alloys. In particular, we have focused on two of the above factors: clustering of SiC particles and matrix microstructure effects. The independent and interactive effects of these factors will be discussed below.

Clustering Effects

The results in Figure 10 indicate that improved combinations of strength and toughness are obtained in the 20% by volume SiC reinforced MMC material with the higher SiC:Al PSR (i.e., larger average SiC particulate size) regardless of matrix temper. Similar results have been observed in MMC's with lower SiC particulate volume fractions (4). This initially seems to conflict with conventional wisdom, which states that the smaller the particle size, the better the damage tolerance at a given strength level (15). In particular, in these high volume fraction particle hardened materials, in which the void nucleation process should have a strong influence on the overall fracture resistance, it has been shown experimentally that larger particles nucleate voids earlier in the deformation process (16). Work by Argon, et al. (17), however, had demonstrated that the effect of particle size on the void nucleation process was not necessarily intrinsic, but rather was a result of the distribution of local volume fractions in high volume fraction microstructures. As the local volume fraction increases, so do the stresses in the region of the closely spaced particles. Thus, void nucleation events such as particle cracking or particle/matrix decohesion can occur at lower levels of deformation in these regions of the microstructure. Quantitative metallographic evaluation of the materials used in this work revealed that the composites containing the smaller SiC particulates (SiC:Al PSR = 0.3:1) exhibited increased clustering compared to the composites containing the larger sized SiC particulates (Figure 7). In addition, metallographic sectioning of fractured tensile specimens revealed that sub-surface cracking associated with clustered regions appeared to be more prevalent in the SiC:Al PSR = 0.3:1 material (Figure 11a) as compared to the less clustered SiC:Al PSR = 0.7:1 materials (Figure 11b). These observations are consistent with the role of local volume fraction on the void nucleation process.

To gain further insight into the role of clustered particles on the fracture initiation process in the presence of a stress concentrator, double notched bend bars of the geometry shown in Figure 16 were tested. These types of specimens have been successful in identifying fracture initiation events in materials which fracture at low macroscopic strains, such as cleavage (18,19) or intergranular (20) fracture, and the results can be analyzed with currently available finite element analyses. In a typical test conducted in four point bending, only one of the notches fails. The remaining notch thereby represents a specimen unloaded prior to catastrophic fracture and can be metallographically sectioned and examined for evidence of fracture initiation events. Figure 17 shows a section of the remaining notch in a specimen of the SiC:Al PSR = 0.3:1 material heat treated to the T6 temper. The arrowed regions indicate the preferential damage near the notch root in regions of clustered particles. Similar features were observed in the underaged and overaged temper. Further efforts to study initiation events will include the use of imposed hydrostatic pressure during tensile testing. This will extend the range of attainable strains and should increase the number of active initiation sites.

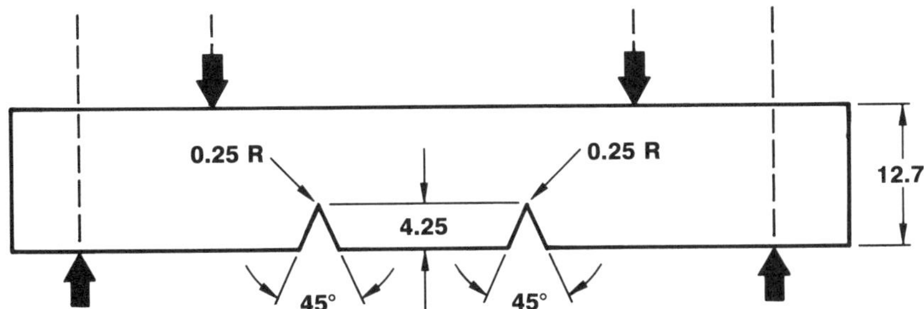

Figure 16 - Geometry of the double notched bend bar used to investigate damage initiation events. Dimensions are in mm.

Figure 17 - Preferential damage initiation (at arrows) in regions of clustered SiC particulates near the notch root of a double notched bend bar. MB78 + 20% by volume F-600 grade SiC particulate, T6 temper.

Matrix Effects

In determining the effects of matrix microstructure on the behavior of composite systems, it must be remembered that the matrix material in a composite may not necessarily deform in a manner similar to that of the unreinforced matrix material. It has been shown in previous work that the aging conditions in unreinforced aluminum alloys strongly affect the deformation characteristics (2,22). For example, unreinforced underaged alloys tend to exhibit serrated flow, shear banding, and flow localization due to the availability of mobile solute and the nature of the coherent particle/dislocation interactions which result in shearing of the aging precipitates. In contrast, overaged microstructures exhibit more uniform deformation due to the tendency for dislocation looping of incoherent particles (23). The underaged composite microstructures tested in the present work failed to exhibit evidence of serrated flow, shear banding, or flow localization. Thus, it is clear that although the composite was heat treated to the underaged condition, the behavior of the matrix is clearly affected by the presence of the SiC particulate. Both the SiC particulate and the substructure would reduce the mean free path for dislocation motion in the matrix, and should therefore reduce the tendency for shear banding. Similar effects have been seen in unreinforced underaged alloys which have been deformed prior to tensile deformation (24) where the dislocation substructure prevents or impedes the passage of shear bands.

In spite of the change in deformation characteristics of the matrix microstructure due to the addition of the SiC particulate, composites in the underaged and overaged conditions still exhibit clear differences in both tensile ductility and fracture toughness behavior. Quantitative fractography on the tensile specimens indicated that overaged (T7X2) specimens exhibited significantly less cracking of the SiC particulates compared to that of the underaged (W2) composites. In addition, the overaged composite exhibited evidence for interface and near interface failure as shown in Figures 13 and 15. The TEM micrograph in Figure 9 shows preferential precipitation at the SiC/matrix interface in an overaged composite, while such precipitation was not observed in the underaged material. TEM foils taken below the tensile fracture surface in the overaged specimens indicated that the particles at the SiC/matrix interface may be preferential sites for void nucleation, as shown in Figure 15. Again, this tendency for interface failure was not observed in the underaged specimen conditions. These results suggest that the role of the interface precipitation is primarily to lower the total interface strength, such that decohesion becomes favored over particle cracking at certain favorable sites during tensile testing.

Work is continuing on quantifying the fractographic features observed in the toughness and notched bend specimens. Preliminary work (26) has indicated that the larger particles are again sampled in preference to the smaller ones in both the W2 and T7X2 tempers and for both specimen types (i.e., notched bend and fracture toughness). Consistent with the fracture toughness data, the notched bend studies similarly revealed that catastrophic fracture occurred at lower levels of stress (and strain) for the overaged specimens. Furthermore, initial fractographic investigations on the fracture toughness specimens also revealed evidence for interface decohesion in the overaged specimens, similar to that seen in the tensile specimens. Accompanying this change in fracture mode in the fracture toughness specimens in going from the underaged to the overaged condition was a corresponding drop in the fracture toughness at a given strength level, as shown earlier (Figure 10). Although not explicitly evaluated in this program, it is possible that the constant low value for the toughness of the composite materials with increased overaging is due to the offsetting effect of increased extent of interface precipitation (reducing toughness) and lower matrix yield strength (increasing toughness).

Quantitative fractography also revealed that the size distribution of the fractured SiC particulates tended to be skewed toward the upper end of the particulate size distribution present in the composite. Similar observations of fracture sampling the larger particles in the distribution were recently reported by Ritter (25). In the present work, this size effect, where the larger particles fracture in preference to the smaller ones, may be due to the greater probability of finding critically sized defects within the larger particles (see Figure 8). Thus, the intrinsic strength of the SiC particulate is considerably less due to the defects present in the larger particles. It is not clear whether these defects were present in the original SiC particulate, or were produced as a result of the processing operations.

Combined Matrix and Clustering Effects

While the individual effects of the matrix microstructure and clustering on properties have been treated separately above, synergistic effects may be particularly important, resulting in a low toughness material. TEM analyses occasionally revealed the presence of voids between closely spaced SiC particulates, Figure 18, as well as accelerated precipitation. In the overaged condition, these regions may be preferential sites for void nucleation through the joint actions of the high local volume fraction of particles, high stresses, precipitation at the SiC/matrix interface, and possible regions of pre-existing voids. However, these voided regions do not appear to be as important for the underaged materials, since tensile fracture in those cases was predominantly by the fracture of the SiC particulate. In the notched bend bar experiments, it was demonstrated that preferential fracture initiation occurred at clustered regions in the composite. Whether this is primarily due to the effects of preferential void formation during processing, preferential precipitation of second phases during aging, elevated local stresses due to clustering, or a combination of these is uncertain at this time. Work is continuing on investigating the fracture nucleation events in the different SiC:Al PSR materials in order to characterize the mechanical properties obtained as a function of clustering, as well as any interactive effects with the matrix microstructure.

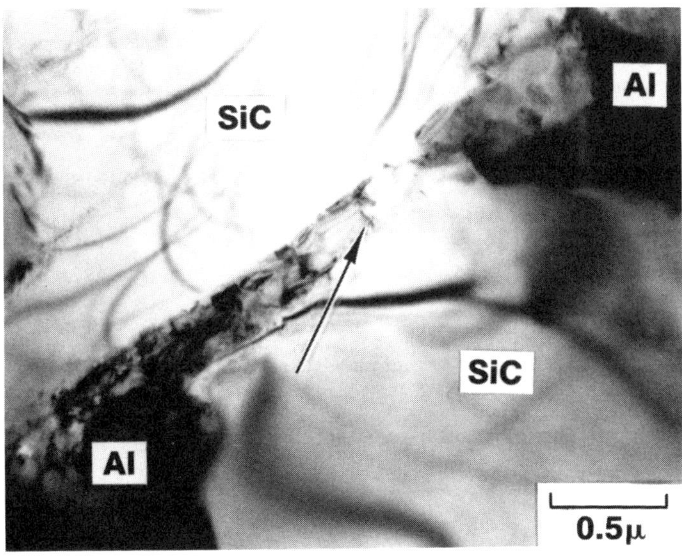

Figure 18 - TEM micrograph of void formation in a region between two closely spaced SiC particulates as a result of processing.

Conclusions

1. The distribution of particle local volume fractions outweighs the effect of SiC particulate size in determining the strength-fracture toughness relationship, regardless of matrix temper. This is probably due to preferential fracture initiation in clustered particle regions.

2. Matrix temper has a strong effect on the fracture toughness, fracture mode, and ductility in tensile specimens. In particular, the primary mode of tensile fracture in underaged materials was SiC particulate fracture, with the larger particles preferentially fracturing and being detected on the fracture surface. In the overaged composite materials, toughness did not recover with increased overaging, as would be expected in the unreinforced matrix alloy. This appears to be due to precipitation at the SiC particulate/matrix interface which reduces the interface strength, and results in an increasing proportion of fracture by particle/matrix decohesion at the expense of particle fracture in tensile specimens.

3. As with the tensile specimens, the notched bend and fracture toughness specimens similarly demonstrated that the larger size particles in the distribution participated in the fracture event, and that particle decohesion was more favored in the overaged specimens. Work is continuing on quantifying the differences in fractography for the underaged and overaged specimens as compared to the tensile specimens in order to study the effect of stress state on the fracture process.

Acknowledgments

This cooperative work is supported by funds provided by Alcoa and DARPA-ONR N00014-86-K-0773. The authors would like to acknowledge the contributions of Celeste Cook, Kevin Armanie, and Thomas Gurganus of Alcoa Laboratories to this work. Permission to publish this paper is appreciated.

References

1. I. Kirman, Metallurgical Transactions, 2A (1971) 1761-1770.

2. P.N.T. Unwin and G. C. Smith, J. Inst. Metals, 96 (1969) 299.

3. G. C. Garrett and J. F. Knott, Metallurgical Transactions, 9A (1978) 1187-1201.

4. W. H. Hunt, Jr., "Fracture Processes in SiC-Al Composites" (Paper presented at 9th Annual Discontinuous Reinforced Metals Working Group Meeting, Park City, UT, 7 January 1987).

5. P. K. Mirchandani, "Structure-Mechanical Property Relations in Aluminum-Ceramic Composites" (Ph.D. thesis, Michigan Technological University, 1987).

6. T. G. Nieh, et. al., "Microstructure and Fracture in SiC Whisker Reinforced 2124 Aluminum Composite," Proceedings, Fifth International Conference on Composite Materials, ed. W. C. Harrigan, Jr., et. al., (Warrendale, PA: The Metallurgical Society, 1985), 825-842.

7. W. A. Spitzig, et. al., Metallography, 18 (1985) 235-261.

8. L. M. Barker and F. I. Barratta, J. Testing and Evaluation, 8(3) (1980) 97-102.

9. R. J. Arsenault, Materials Science and Engineering, 64 (1984) 171-181.

10. S. V. Nair, et. al., Int. Met. Rev., 30 (1985) 275-290.

11. C. P. You, et. al., Scripta Met., 21 (1987) 181-187.

12. C. R. Crowe, et. al., "Microstructure Controlled Fracture Toughness of SiC/Al Metal Matrix Composites," Proceedings, Fifth International Conference on Composite Materials, ed. W. C. Harrigan, Jr., et. al., (Warrendale, PA: The Metallurgical Society, 1985), 843-866.

13. R. Van Stone, et. al., Int. Met. Rev., 30 (1985) 157-179.

14. J. D. Embury, Metallurgical Transactions, 16A (1985) 2191-2200.

15. D. Broek, Engineering Fracture Mechanics, 5 (1973) 55-66.

16. J. R. Fisher and J. Gurland, Met. Sci., (1981) 185-192.

17. A. S. Argon, et. al., Metallurgical Transactions, 6A, (1975) 825-837.

18. J. J. Lewandowski and A. W. Thompson, Proc. ICF6, ed. S. R. Valluri, et. al., 2 (1986) 1985-1995.

19. J. J. Lewandowski and A. W. Thompson, Metallurgical Transactions, 17A (1986) 1769-1786.

20. J. J. Lewandowski, et. al., Acta Met., 35 (1987) 593-609.

21. J. P. Griffiths and D. R. J. Owen, J. Mech. Phys. Solids, 19 (1971) 419-431.

22. J. J. Lewandowski and J. F. Knott, Proc. ISCMA-7, ed. H. J. McQueen, et. al., 2 (Pergamon Press, 1985), 1193.

23. J. W. Martin, Micromechanisms in Particle Hardened Alloys, (Cambridge University Press, 1980).

24. D. J. Lloyd, (Paper presented at Fall TMS-AIME Meeting, Orlando, FL, October 1986).

25. A. M. Ritter, et.al., this volume.

26. C. Liu and J. J. Lewandowski, unpublished results, Case Western Reserve University, 1987.

SUPERPLASTIC CHARACTERISTICS OF MODIFIED 7075 Al AND Al - Ti ALLOYS PROCESSED BY RAPID SOLIDIFICATION POWDER METALLURGY

G.S. Murty* and M.J. Koczak

Department of Materials Engineering
Drexel University
Philadelphia, PA 19104

*On leave from Indian Institute of Technology, Kanpur, India

Abstract

The high temperature deformation behavior of Al - Zn - Mg alloys with additions of Mn and Mn / Si, and Al - Ti alloys processed by rapid solidification powder metallurgy route was investigated in order to explore their superplastic characteristics. A wide variation in the deformation response of these alloys was noticed despite their microstructural similarity. A high rate sensitivity of flow stress and extended ductility were observed in the Al - Zn - Mg - Mn and Al - 4% Ti alloys, whereas the other alloys exhibited normal ductility. The variations in their deformation response are interpreted in terms of a threshold stress for superplastic flow, that depends on the particle size, volume fraction and the nature of the dispersoid.

Processing and Properties for Powder Metallurgy Composites
Edited by P. Kumar, K. Vedula and A. Ritter
The Metallurgical Society, 1988

Introduction

Rapid solidification powder metallurgy processing route offers unique advantages in achieving fine scale dispersions of hard particles in structural Al alloys [1]. While their room temperature strength increase is accompanied by a significant drop in ductility, a high elevated temperature ductility may be expected because of their fine grain size. One of the requirements for exhibiting superplasticity is a fine grain size that is stable at elevated temperatures. Any improvement in their elevated temperature ductility is of interest in connection with their deformation processing. The observations on superplasticity and extended ductility of powder metallurgy Al alloys were recently reviewed by Wadsworth et al .[2,3]. Our recent work [4,5] on some rapid solidification processed and mechanically alloyed Al alloys indicated extended ductility in some alloys, while others exhibit normal ductility. The observations on the variations in the high temperature deformation response of modified 7075 - Al and Al - Ti alloys are presented with interpretations in terms of the superplastic flow mechanisms.

Experimental Details

The high temperature deformation response of two groups of alloys was investigated. One category is the 7XXX series Al alloys modified with the additions of Mn and Mn / Si in order to improve their room temperature strength and modulus of elasticity. The prealloyed compositions were helium atomized by Valimet, Stockton, CA. The powder was canned, degassed at 330°C and extruded with a reduction ratio of 16:1 in the temperature range of 350 - 450°C by Nuclear Metals, Concord, MA. Further details of these alloys are available elsewhere [6].

The second group is Al - Ti alloys containing fine dispersions of thermally stable phases. The prealloyed powder prepared by helium atomization was consolidated through vacuum hot pressing at 493°C and extruded at 400°C with a reduction ratio of 47:1. The material processed in this manner is designated as AT alloys. In one case, the prealloyed powder prepared by helium atomization was mechanically alloyed and subsequently processed as above. This alloy designated as AM has the additional dispersoids of Al_4C_3 and Al_2O_3 as a result of mechanical alloying. The details of alloy compositions, dispersoids and their volume fractions estimated from stoichiometry in both the groups of alloys are listed in Table 1.

Tensile specimens of 15 - 20 mm gage length and 5 mm diameter and compression specimens of 10 mm height and 7 mm diameter were obtained from the extruded bars and tested at elevated temperatures on an Instron machine. An elliptical radiant furnace was employed for attaining the test temperatures. The true stress - true strain rate ($\dot{\varepsilon}$) characteristics and the strain rate sensitivity index ($m = d \log \sigma / d \log \dot{\varepsilon}$) were assessed by the strain rate change test. In some tests, the strain rate change test was repeated on the same sample following the first sequence of crosshead speed changes. The data obtained on a given specimen from the first and second sequence of crosshead speed changes are referred to as first (#1 CY) and second (#2 CY) cycle data. Constant crosshead speed tensile tests were performed to evaluate tensile ductility. Metallographic observations were carried out by transmission electron microscopy (TEM) with a JEM - 100 CX microscope. The thin foils for TEM observations were prepared by twin jet polishing with an electrolyte of 1:3 nitric acid: methanol at -30°C. Le Mont analysis of the Scanning electron microscope image in secondary electron mode was carried out to determine the size distribution of the dispersoids in the alloys.

Table 1. The alloy compositions and details of dispersoids

Alloy	Dispersiod	Volume fraction
Modified 7075 - Al alloys		
1. 7075 - Mn (composition in wt %) Zn: 6.48, Mg: 2.71, Cu: 1.95 Mn: 3.87, Fe: 0.16, Cr: 0.21 Al: Bal	$MnAl_6$	0.15
2. 7075 - Mn / Si Zn: 8.00, Mg: 1.75, Cu: 1.95 Mn: 3.80, Si: 1.00, Al: Bal	$Mn_3Si_2Al_{15}$	0.15
Al - Ti alloys		
3. AT - 4 Al - 4 wt% Ti	Al_3Ti	0.10
4. AT - 6 Al - 6 wt% Ti	Al_3Ti	0.15
5. AM - 4 Al - 4 wt% Ti Mechanically alloyed	Al_3Ti $Al_4C_3 + Al_2O_3$	0.10 0.04

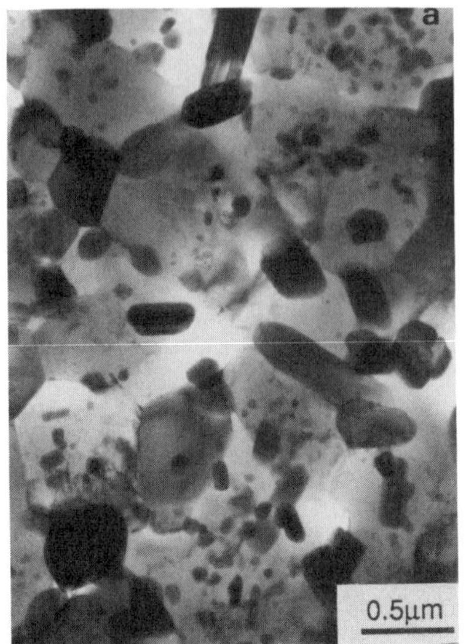

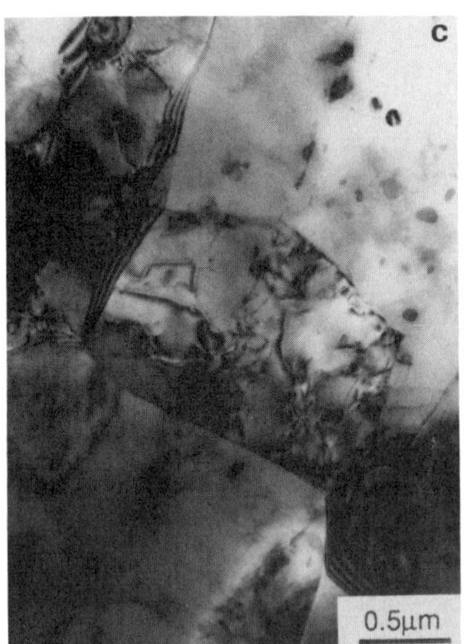

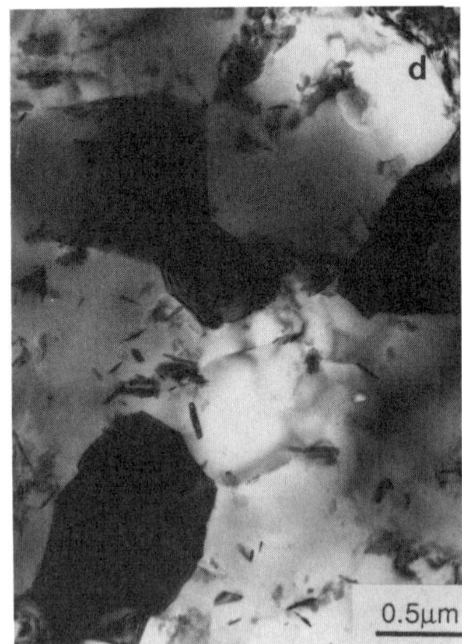

Figure 1. Electron micrographs of alloys: (a) 7075 - Mn (b) 7075 - Mn / Si (c) Al - 4% Ti (d) Al - 4% Ti (mechanically alloyed)

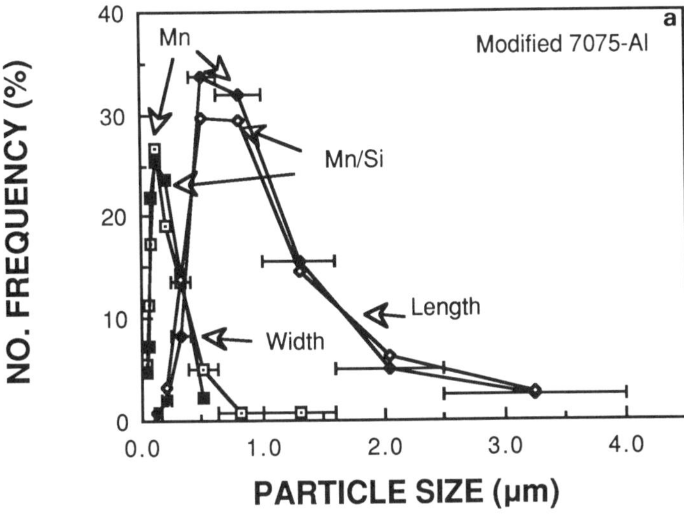

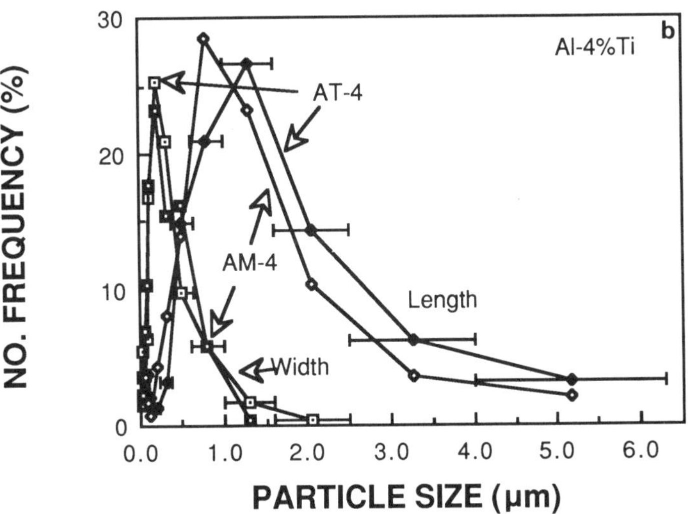

Figure 2. Le Mont analysis of the dispersoid particle size distribution: (a) Modified 7075 (b) Al - 4% Ti

Results

As a result of the rapid solidification powder metallurgy route adopted in their processing, fine equiaxed grains with a mean diameter of ~ 1 - 2 μm were observed in these alloys. Furthermore, the spotty rings characteristic of the matrix grains, observed in electron diffraction patterns suggest the grains to be having a large misorientation. The electron micrographs representing their microstructures are shown in Fig. 1. Most of the particles being in the form of platelets, their width and length distributions are presented in Fig. 2. However, there are some particles of finer size than the data shown in Fig.2 as indicated byTEM observations. On exposure to temperatures in the range 520 - 575°C for 120 hours, the coarsening of grain size was less than a factor of two in these alloys.

Despite their similarity in microstructures, the high temperature deformation characteristics of the various alloys are quite different. In terms of the rate sensitivity of flow stress, the alloys can be grouped into two distinct types. The alloys 7075 - Mn and AT - 4 exhibit a relatively high strain rate sensitivity of flow stress, which is a characteristic of superplastic materials. The other alloys (7075 - Mn / Si, AT - 6 and AM - 4) are of lower rate sensitivity typical of normal plasticity. These differences in their high temperature deformation behavior can be noted from the stress - strain rate and m - strain rate plots shown in Figs. 3 - 5. The maximum rate sensitivity is seen to occur at a relatively high strain rate of ~ 10^{-1} / s in all cases. From the plots shown in Figs. 3 - 5, the second cycle data of the strain rate change test indicate a drop in flow stress from that of the first cycle and it is accompanied by an increase in the rate sensitivity index. From the microstructural observations before and after the strain rate change test, grain coarsening by a factor of ~ 1.5 was observed as a result of the deformation accumulated during the first cycle of the strain rate change test. In view of this tendency of improved rate sensitivity coupled with grain coarsening, some of the specimens were coarsened by a long anneal of ~ 120 hours in the temperature range of 520 - 575°C and subsequently tested. The effect of coarsening on the deformation characteristics shown in Figs. 6 and 7 is similar to that of prestraining by the strain rate change test.

The elevated temperature tensile ductility of these alloys are summarized in Table 2. In accordance with their better rate sensitivity of flow stress, extended ductility was observed in 7075 - Mn and AT - 4 alloys. The ductility of the coarsened 7075 -Mn alloy was , however, low despite its higher rate sensitivity of flow stress and it is limited by cavitation. Longitudinal metallographic sections of the fractured tensile specimens were examined to assess the extent of cavitation in various alloys. Extensive cavitation damage was noted in all alloys except AT - 4, as shown in Fig. 8.

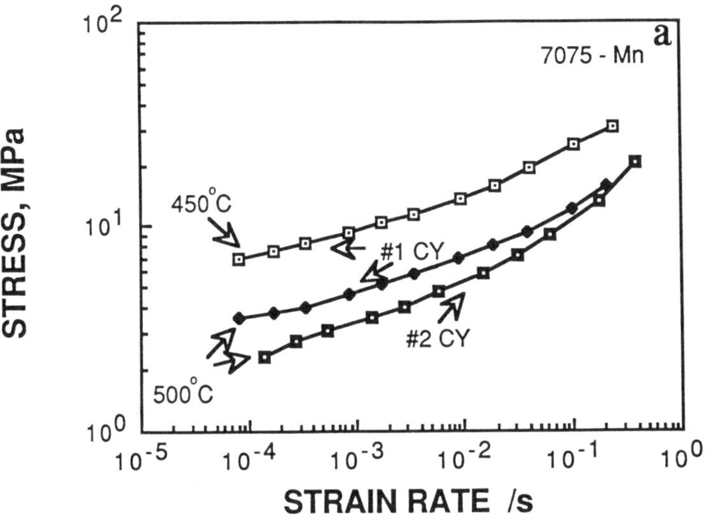

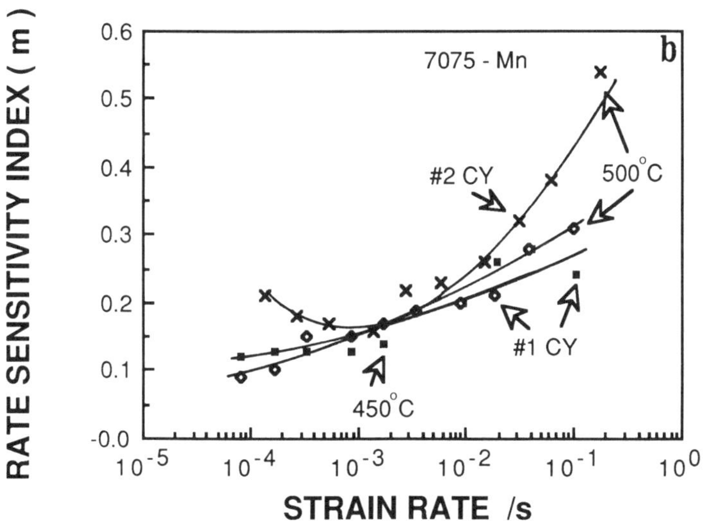

Figure 3. (a) σ - $\dot{\varepsilon}$ and (b) m - $\dot{\varepsilon}$ plots for the 7075 - Mn alloy

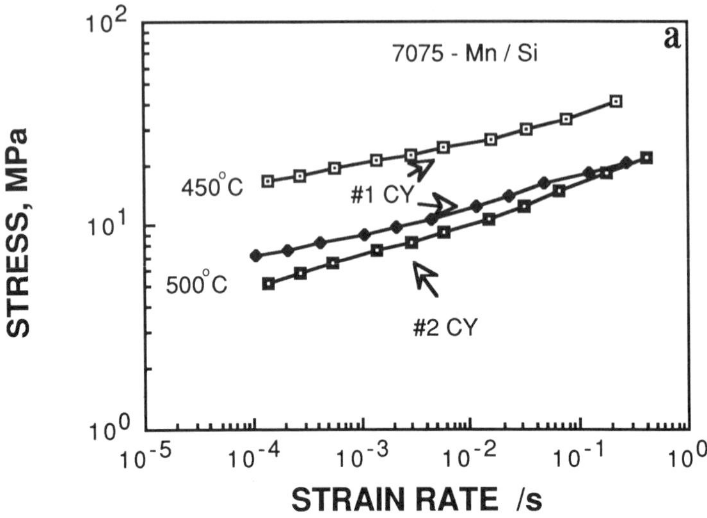

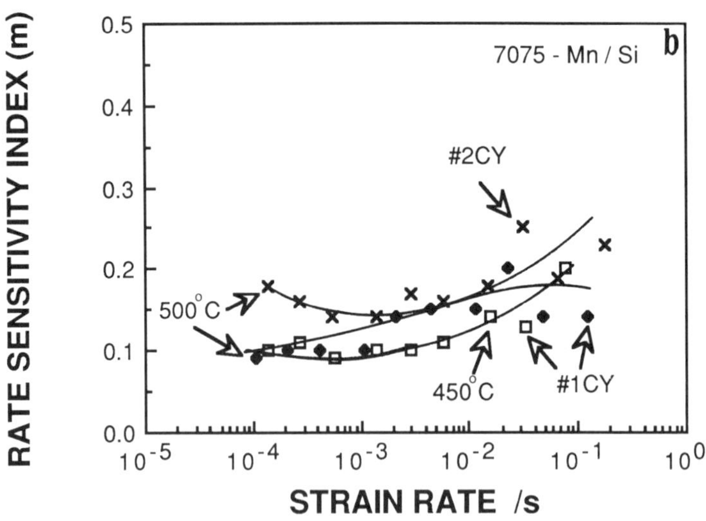

Figure 4. (a) σ - $\dot{\varepsilon}$ and (b) m - $\dot{\varepsilon}$ plots for the 7075 - Mn / Si alloy

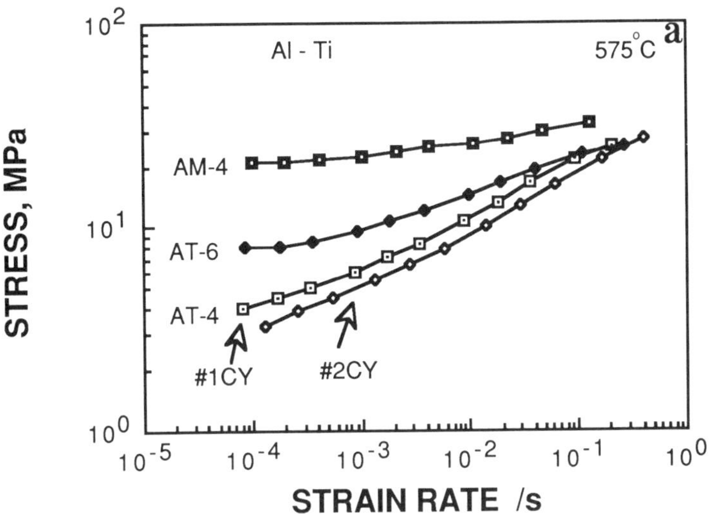

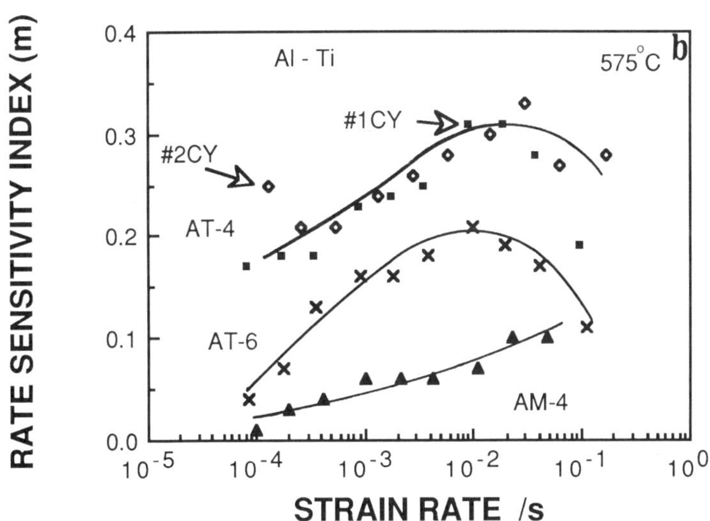

Figure 5. (a) σ - $\dot{\varepsilon}$ and (b) m - $\dot{\varepsilon}$ plots for the Al - Ti alloys

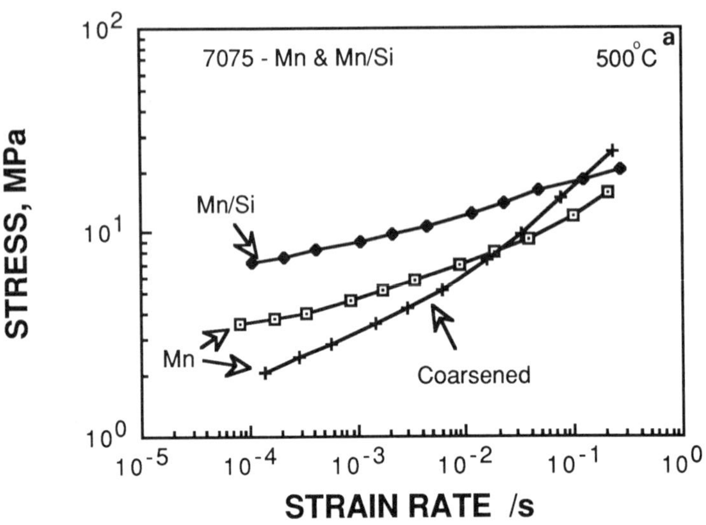

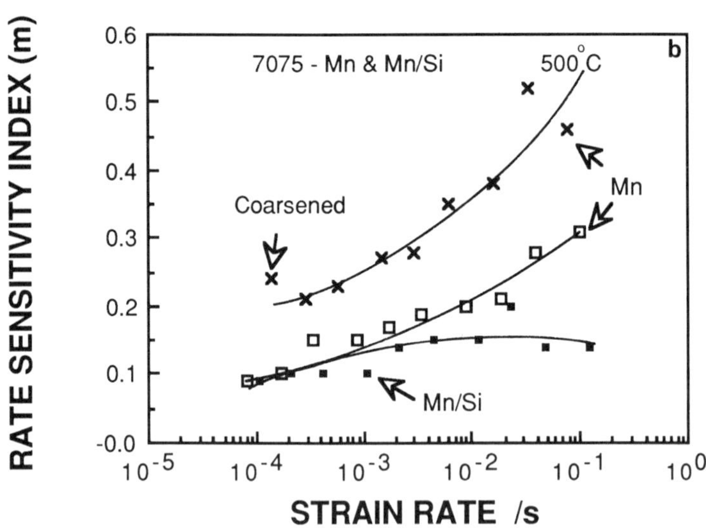

Figure 6. (a) σ - $\dot{\varepsilon}$ and (b) m - $\dot{\varepsilon}$ plots at 500°C for the modified 7075 - Al alloys

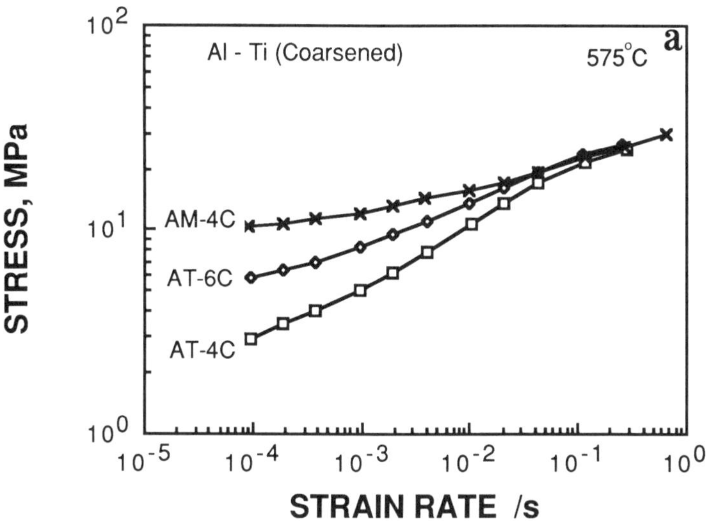

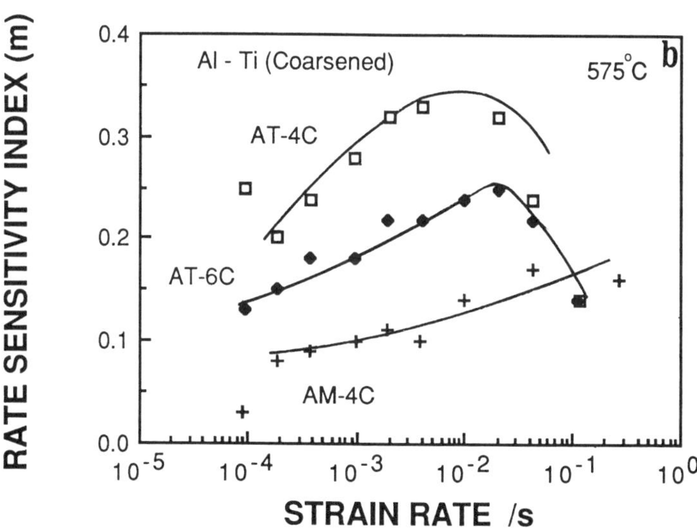

Figure 7. (a) σ - $\dot{\varepsilon}$ and (b) m - $\dot{\varepsilon}$ plots at 575°C for the coarsened Al - Ti alloys

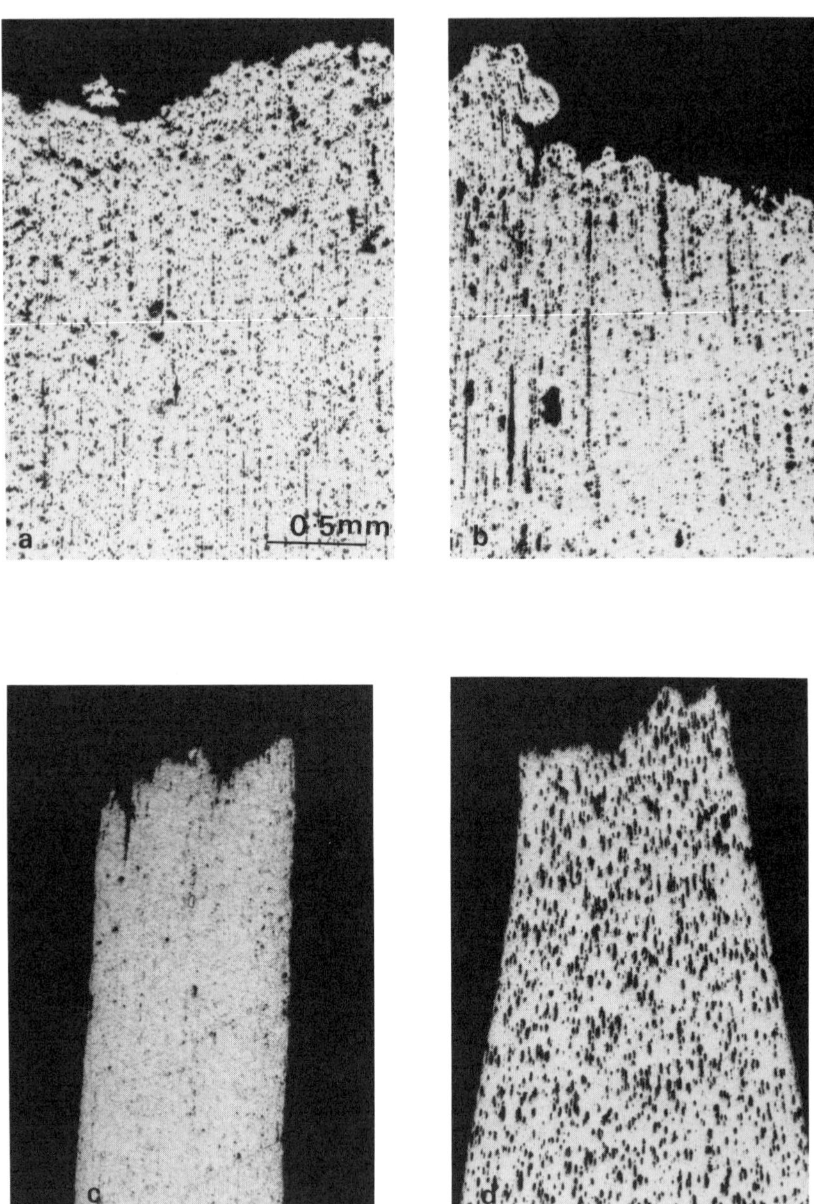

Figure 8. Photomicrographs of fracture zone indicating cavitation in (a) 7075 - Mn (475°C) (b) 7075 - Mn / Si (525°C) (c) AT - 4 (575°C) (d) AM - 4 (575°C) [$\dot{\varepsilon}_{initial} \approx 0.1\ s^{-1}$ in all cases].

Table 2. Tensile ductility data

Alloy	Grain Size (μm)	% elongation	% Reduction in Area	m	T (°C)
Modified 7075 Al alloy					
7075 - Mn	1.5	200	82	0.30	475
7075 - Mn (coarsened)	2.5	80	61	0.40	450
7075 - Mn / Si	1.6	71	64	0.20	525
Al - Ti alloys					
AT - 4	1.6	307	94	0.30	575
AT - 4 (coarsened)	2.6	340	95	0.35	575
AT - 6	1.6	112	70	0.20	575
AT - 6 (coarsened)	1.8	130	77	0.25	575
AM - 4	0.8	85	83	0.10	575
AM - 4 (coarsened)	1.0	98	95	0.15	575

Discussion

On the basis of the fine grain size of these alloys, the same deformation mechanism is likely to dominate under similar test conditions and consequently superplastic behavior is expected over some strain rate range in all the alloys. But the results indicate distinct differences in their deformation behavior. Despite the variation in m among different alloys, two aspects of similarities in their behavior are noteworthy. These are (i) an increase in m with increasing strain rate and (ii) a drop in flow stress acompanied by an increase in m as the microstructure is coarsened. The increase in m with increasing strain rate is typical of the transition from region 1 to region 2 of superplastic materials exhibiting stress - strain rate curves of sigmoidal shape in general. The sigmoidal nature of the stress - strain rate plots and the microstructural coarsening effects support the presence of a threshold stress for the operative mechanism of superplastic flow. Since the grain size in these alloys is stabilized by Zener pinning of the boundaries by second phase particles, the grain and particle coarsening processes are strongly coupled. A threshold stress that varies inversely with grain size can account for the observed microstructural coarsening effects. Consequently, the variations in the response of different alloys can be interpreted in terms of wide differences in the magnitude of the threshold stress. As the threshold stress for superplastic flow increases, its effect in lowering the rate sensitivity of flow stress continues up to a high strain rate. The lower flow stress of the 7075 - Mn and AT - 4 alloys at low strain rates suggests a lower threshold stress for superplastic flow in these alloys relative to the other alloys of this study.

The possible interpretations of the differences in the magnitude of the threshold stress among these alloys will now be examined in terms of the mechanisms for its origin in superplastic flow. A threshold stress for superplastic flow arises because of a barrier in driving an essential event of the operative mechanism. It is well known that grain boundary sliding (GBS) with an appropriate accommodation is the dominant mode of superplastic deformation. Considering Ashby - Verrall model [7] involving diffusionally accommodated GBS as the basis for discussion, several distinct processes may contribute to the threshold stress for superplastic flow, as listed below:

1. An increase in the grain boundary area during the grain neighbor switching process.
2. An interfacial barrier for diffusional accommodation since the ability of grain boundaries to act as sinks and sources for point defects is affected by the particles at the grain boundaries [8].
3. The inhibition of GBS by the dispersed particles at the grain boundaries.
4. The inhibition of grain boundary migration by particles (Zener pinning).

While the predicted threshold stress from the increase in the grain boundary area during the grain neighbor switching process is too small, all the other processes can contribute to a significant threshold stress, whose magnitude increases with the finer particle size and volume fraction of the dispersoid.

Among the Al - Ti alloys, a high threshold stress in the mechanically alloyed condition arises from the fine dispersion of Al_4C_3 and Al_2O_3 particles and the variations in the rate sensitivity of flow stress can be understood on the basis of considerations cited above. But in the other group, the higher threshold stress of the 7075 - Mn / Si alloy over that of the 7075 - Mn cannot be rationalized in terms of the particle size and volume fraction of the dispersoids. Consequently, the variation in threshold stress may arise from the consideration of additional factors such as the nature of the particle / matrix interface, the atom mobility within the dispersoid phase, the hardness of the particle etc. A quantitative prediction of the threshold stress for superplastic flow may thus have to be based on the inherent nature of the particle apart from its size and volume fraction.

This study further suggests that the rate sensitivity of flow stress of all these alloys can be improved through an appropriate thermomechanical processing that lowers the threshold stress for superplastic flow by coarsening the microstructure. The improved rate sensitivity may, however, not lead to better tensile ductility in some alloys because of cavitation, as seen from the observations on the coarsened 7075 - Mn alloy. In alloy design for superplasticity coupled with the optimum properties for structural applications, the significant factors that must be considered are the volume fraction, size scale and nature of particles.

Conclusions

1. Among the modified 7075 - Al alloys, high rate sensitivity of flow stress and extended ductility are observed in the 7075 - Mn alloy. On the other hand, the 7075 - Mn / Si alloy exhibits normal ductility despite the fine grain size stabilization by particles. In both these alloys, extensive cavitation occurs and limits the level of tensile ductility.

2. In the Al - Ti alloys, AT - 4 exhibits high rate sensitivity and extended ductility, whereas only normal ductility with a low rate sensitivity is observed in the mechanically alloyed AM - 4. Extensive cavitation occurs in the AM-4 alloy and it is minimal in the AT - 4 alloy.

3. The variations in the deformation response of the alloys are interpreted in terms of a threshold stress for superplastic flow, which depends on the volume fraction, particle size and the nature of the dispersoid.

Acknowledgements

This work was supported by the Naval Air Development Center, Warminster, PA. The authors would like to thank Drs. U.V. Deshmukh and W.E. Frazier, who provided the alloys for this investigation.

References

1. I.B. MacCormack, "Dispersion strengthened P/M aluminum alloys" Metals and Materials, 2 (1986) 131 - 137.

2. J. Wadsworth et al. in "High Strength powder metallurgy Aluminum alloys II", eds. G.J. Hildeman and M.J. Koczak (Warrendale, PA: The Metallurgical Society, 1986) 137 - 154.

3. T.G. Nieh and J. Wadsworth, in "Superplasticity in Aerospace - Aluminum", eds. R. Pearce and L. Kelly (Cranfield, England, 1986) 194 - 214.

4. G.S. Murty and M.J. Koczak, "Superplastic Characteristics of modified 7075 Aluminum alloys processed through rapid solidification route", Scripta Metallurgica, 20 (1986) 1721 - 1726.

5. G.S. Murty, M.J. Koczak and W.E. Frazier, "High temperature deformation of rapid solidification processed / mechanically alloyed Al - Ti alloys", Scripta Metallurgica, 21 (1987) 141 - 146.

6. M.J. Koczak and U.V. Deshmukh, "Development of high modulus corrosion resistant aluminum alloys", (Final Report NADC - 86074 - 60, Naval Air Development Center, 1986).

7. M.F. Ashby and R.A. Verrall, "Diffusion - Accommodated flow and Superplasticity", Acta Metallurgica, 21 (1973) 149 - 163.

8. E. Arzt, M.F. Ashby and R.A. Verrall, "Interface controlled diffusional creep", Acta Metallurgica, 31 (1983) 1977 - 1989.

LIGHT-WEIGHT, HIGH-PERFORMANCE METAL MATRIX COMPOSITES

H. J. Rack

Department of Mechanical Engineering
Clemson University
Clemson, South Carolina

Abstract

Recent years have seen the development of a wide range of light-weight, high-performance aluminum powder metallurgy composites. These materials have combined both standard, e.g., 6061 and 2124, and specialty matrix compositions, e.g., Al-Mg-Cu-Li and Al-Fe-Ce, with a wide variety of discontinuous reinforcements, e.g., Al_2O_3, B_4C, and SiC. This paper reviews the fabrication and damage tolerance performance of these light-weight, high-performance composites. Particular attention is given to developing a general framework for understanding the interrelationship existing between microstructure, thermomechanical processing and ductility/fracture toughness behavior in these composite systems.

Processing and Properties for Powder Metallurgy Composites
Edited by P. Kumar, K. Vedula and A. Ritter
The Metallurgical Society, 1988

Introduction

Modern design procedures continually strive to increase structural efficiencies through reductions in either absolute weight or increases in the strength-to-weight ratio. Figure 1 illustrates how, for a cargo-bomber aircraft application (1), reductions in material density or increases in modulus (stiffness), yield strength and ultimate tensile strength can all be directly translated to reductions in structural weight. For example, a 10 percent reduction in material density, which might be achieved through substitution of Al-Li alloys for 2000 series aluminum alloys, would result in a 10 percent reduction in structural weight. Alternatively, a 50 percent increase in modulus, which can be achieved through substitution of a silicon carbide (SiC) reinforced aluminum alloy for an unreinforced wrought aluminum alloy (2), would result in a 10 percent reduction in structural weight. Indeed, it is possible to envision the development of a reinforced Al-Li alloy alloy having a density 10 percent less than present wrought high-strength aluminum alloys, with a 50 percent higher stiffness (3,4).

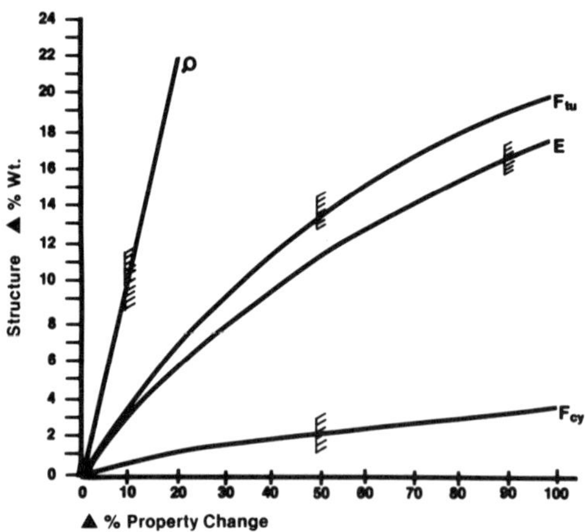

Figure 1 Performance enhancement related structural weight reductions in cargo-bomber aircraft applications (1).

System trade-studies, such as described above, have been the primary motivating factor in the renewed interest shown in metal matrix composites. Historically, wide spread industrial application of these composites has however been limited by the relatively high costs of both the reinforcement fiber and the necessary metal matrix component fabrication procedure. This barrier appears to have been recently broken through the introduction of low cost discontinuously reinforced metal matrix composites (5). This paper will review the production of powder-based, discontinuously reinforced aluminum metal matrix composites. It then considers how alloy modification and appropriate thermomechanical processing procedures can be utilized to enhance the tensile ductility and damage tolerance behavior of these composites.

Fabrication of Discontinuously Reinforced Metal Matrix Composites

A generalized flow chart illustrating the procedures typically utilized during fabrication of powder metallurgy based, discontinuously reinforced, metal matrix composites is presented in Figure 2.

Ceramic Reinforcement
↓
Deagglomeration
↓
Metal Powder -------------→ Mixing
↓
Billet Consolidation
↓
Scalping/Q.A.
↓
Primary Processing
(Rolling, Extrusion, Forging)
↓
Secondary Processing
(Forming, Joining, Machining)

Figure 2 Fabrication sequence for P-M discontinuously reinforced metal matrix composites.

The initial step in the manufacturing sequence involves proper selection of the discontinuous reinforcement and the matrix alloy. Ceramic reinforcements which have been considered for discontinuous metal matrix composite fabrication include SiC and Si_3N_4 whiskers, SiC, B_4C and TiB_2 particulates, and Al_2O_3/Al_2O_3-SiO_2 short fibers, Figure 3. Selection of the appropriate reinforcement should been based upon reinforcement properties, cost, size and matrix-fiber compatibility. For example, a wide range of shapes and sizes are available for both α and β SiC reinforcements. Shapes considered for metal matrix composites have included whiskers, oblate platelets, typically referred to as particulates, and spherical particles. Mean particulate sizes between 3 and 30 μm have also been investigated.

Originally, matrix compositions were limited to available pre-alloyed air or inert gas atomized powders of wrought aluminum alloy compositions, e.g., 6061 and 2124 (5,6). However several studies (7-9) have shown that elements commonly included in wrought aluminum alloys as grain refiners, e.g., Mn and Cr, are unnecessary in discontinuously reinforced metal matrix composites, and indeed may be deterimental to their tensile ductility and damage tolerance properties. Finally, microstructural-mechanical behavior examinations (7,8) of these alloys has shown that leaner alloy compositions tend to develop a better combination of strength, ductility and toughness.

Recently (8,10) newer matrix compositions have also started to be investigated, e.g., 7090, 7091, CU 78, Al-Cu-Mg-Li, IN 952, where the full benefits of rapid solidification or mechanical alloying can be recognized in the discontinuously reinforced metal matrix composite.

Dry or wet blending follow selection of the reinforcement and matrix powder[a]. Figure 4 shows a typical SiC whisker/aluminum powder blend. This figure illustrates an important characteristic of many blends, that is the difference in size distribution between the reinforcement and aluminum matrix powder. For example, the SiC whiskers shown in Figure 4 have a mean diameter of 0.5 μm, with lengths ranging between 5 and 40 μm. In contrast

[a]The blending step may initially require prior diagglomeration of the reinforcement, as for example in the blending of SiC whiskers (11).

Figure 3-Scanning election micrographs of (a) SiC whiskers, (b) Si_3N_4 whiskers, (c) SiC particulates (d) SiC spherical particles, (e) B_4C particulates, (f) TiB_2 particulates and (g) $A\ell_2O_3$ short fibers.

(e)

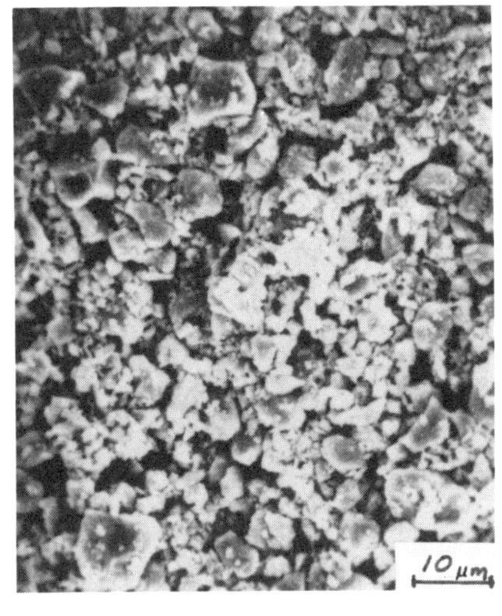

(f)

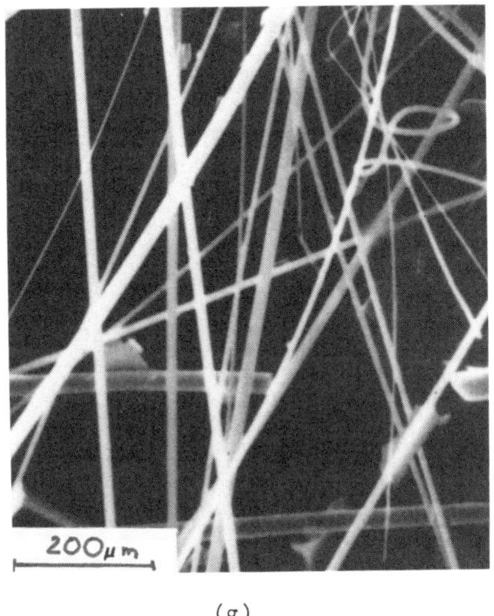

(g)

Figure 4 Scanning electron micrograph of silicon carbide whisker - aluminum blend

the aluminum powder is nominally -325 mesh, with a mean size of 15 μm (8). Various authors' have shown that this difference can lead to whisker clumping, early crack initiation and non-uniform precipitation (8,12-14). However, improved mechanical working procedures can minimize these detrimental effects (15), the primary purpose of the necessary hot/cold deformation being to enhance reinforcement-matrix mixing.

Final billet fabrication involves cold compaction and hot isostactic or vacuum hot pressing, Figure 5. Cold compaction densities should be controlled to maintain open, interconnecting porosity. The latter is extemely important during the outgassing stage of the pressing operation. While the details of reinforcement-powder blend outgassing are generally considered proprietary by the composite manufacturer, it normally involves removal of adsorbed or chemically bound water and other volatile species through the combined action of heat, vacuum and inert gas flushing. For example, outgassing of SiC reinforced aluminum metal matrix composites involves removal of adsorbed water from both SiC and aluminum, as well as chemically bound water from the aluminum alloy. The principal reactions occurring during this outgassing process are

$$Al_2O_3 \cdot 3H_2O \xrightarrow{>100^\circ C} Al_2O_3 \cdot H_2O + H_2O(g)$$

$$Al_2O_3 \cdot H_2O \xrightarrow{>400^\circ C} Al_2O_3 + H_2O(g)$$

$$2Al + 3H_2O \xrightarrow{>200^\circ C} Al_2O_3 + 3H_2(g)$$

$$2Al + 6H_2O \longrightarrow Al_2O_3 \cdot 3H_2O + 3H_2(g)$$

where H_2 and H_2O are the primary gaseous reaction products and Al_2O_3 is the primary solid product.

Once the desired isothermal pressing temperature is reached, final consolidation is accomplished by pressure application. Selection of the consolidation temperature is typically based on the need to minimize the pressures necessary for complete consolidation without degrading the powder

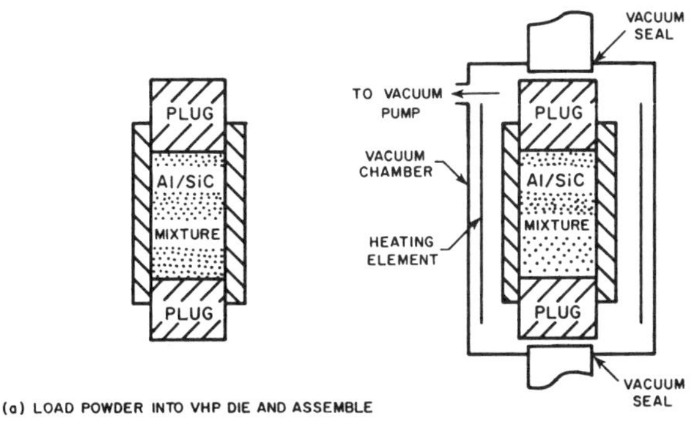

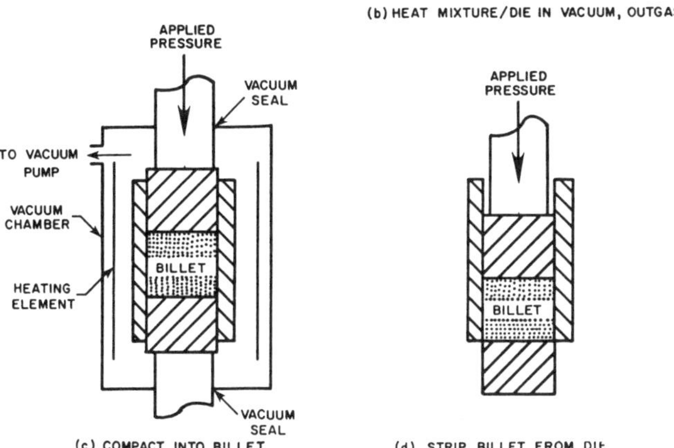

Figure 5 Schematic representation of vacuum hot pressing operation.

matrix. Both solid state and mushy zone techniques have been employed, with some evidence suggesting that higher tensile ductilities may be obtained following solid state pressing (16).

Consolidated billets, typically 98+ percent theoretical density, can be fabricated into a wide variety of shapes utilizing standard metal working equipment. Primary working operations involving rolling, extrusion and forging have all been demonstrated (17-19). Secondary procedures have included shear spinning, superplastic forming and joining (20-22). Each procedure must be adjusted, recognizing the forming temperatures, rates and flow conditions uniquely identified with each composite system. For example, control of short fiber/whisker length-to-diameter ratio and orientation is essential if the full benefits of this composite system is to be realized.

Short fiber length-to-diameter ratio and orientation in extruded products may be controlled through proper choice of die design, extrusion ratio, strain rate and deformation temperature. Figure 6 schematically compares two extrusion die geometries currently employed for discontinuously reinforced aluminum metal matrix composites (18). Figure 6(a) shows the flow fields associated with a shear faced die, as commonly used in the U.S. wrought aluminum industry, while Figure 6(b) shows a "stream-line" flow die. The latter die configuration has been designed to simulate hydrostatic flow conditions, eliminating re-entrant corners and "dead" zones where short fiber reinforcements could undergo sharp velocity

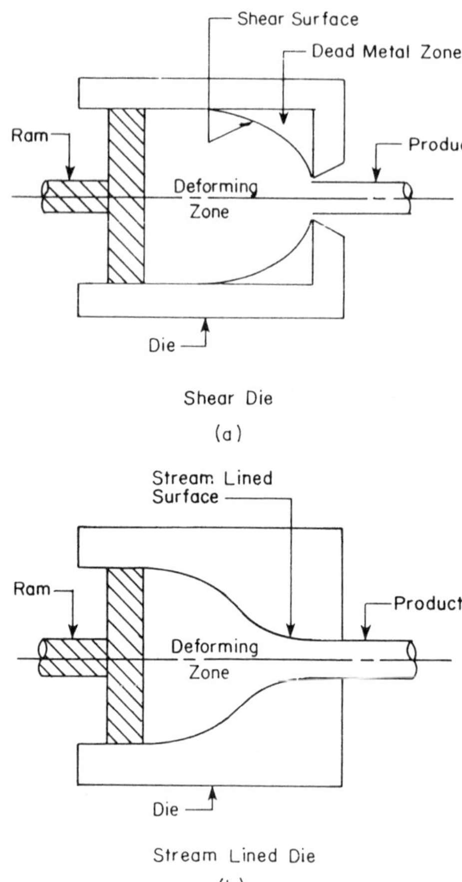

Figure 6 Comparison of (a) shear and (b) streamline flow extrusion die configurations.

Table I. Whisker Aspect Ratio as a Function of Processing (23)

Material	L/D
Powder/whisker blend	19.8
36:1 Extrusion Ratio	
Round-to-Round Through	18.0
Streamline Flow Die	

discontinuities. Table I shows that when this die design approach is used for extrusion of 2124 containing 20 volume percent SiC whiskers minimum whisker damage occurrs.

Several methods of selecting the appropriate strain rates and deformation temperatures for discontinuously reinforced metal matrix composites have been suggested. Most are based on experience or trial and error. One approach which appears to warrant further consideration utilizes the material's true stress-true strain rate constitutive behavior as a function of strain rate and temperature (24). If the dynamic constitutive behavior is represented by a relationship of the form,

$$\sigma = A\dot{\varepsilon}^m$$

then the efficiency, ζ, of the deformation conditions, that is the amount of energy transformed into shape change, can be defined by

$$\zeta = \frac{2m}{m + 1}$$

where m, is a function of both temperature and strain rate.

Figure 7 illustrates application of this technique to 2124 reinforced with 20 volume percent silicon carbide whiskers (25). This data suggests that the maximum amount of useful work can be obtained when deformation processing of this composite is carried out at 485°C and 10^{-4} sec^{-1}.

A word of caution should be presented at this juncture. Maximum useful work, as defined above, need not be associated with optimal properties. For example, Gegel et al (25) have shown that this maxima in 2124 reinforced with 20 volume percent SiC whiskers, is associated with nearly complete dynamic recovery, higher temperatures and rates leading to incipient melting, lower temperatures and rates to dislocation accumulation. In contrast, it is now well known however that maximum toughness in wrought aluminum alloys is generally associated with an unrecrystallized grain structure. Future enhancements in the toughness of extruded SiC whisker reinforced aluminum are therefore expected to require extrusion at temperatures below "optimal".

Damage Tolerance Behavior of Discontinuously Reinforced Metal Matrix Composites

Several recent investigations have focused on the ductility and fracture toughness behavior of discontinuously reinforced metal matrix composites(7-10,27-33). These studies suggest that differing factors control fracture initiation and propagation in these materials. It appears that the most important factor currently limiting the tensile ductility of discontinuously reinforced aluminum matrix composites is the presence of large inclusions, typically SiO_2, oversized SiC particulates or impurity related aluminum intermetallics, e.g., $Al_{20}Cu_2(Cr,Fe)_3$ in SiC reinforced 2124 (7,35). Elimination of these inclusions has been shown to increase the ductility of reinforced 2124 aluminum from 2 to 5 percent (34). Further enhancements in ductility can be expected by omitting normal grain size refinement alloying additions, e. g., Mn in 2124, and by avoiding consolidation above the solidus temperature.

Ultimately the ductility of SiC reinforced composites will be limited by the reinforcement. Nutt (29) has suggested that in SiC whisker reinforced aluminum, void initiation will occur by strain localization at the end of the whisker, as depicted in Figure 8. Void formation will occur

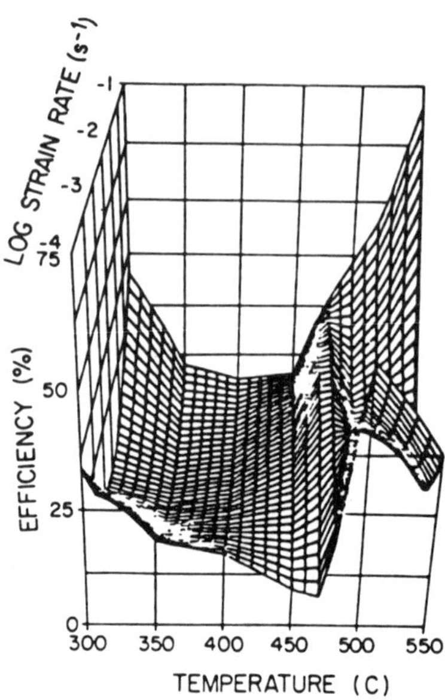

Figure 7 Influence of temperature and strain rate on the deformation efficiency.

at the interface corners between the whisker end and the matrix, with void growth progressing across the whisker end. A similar void initiation mechanism might be envisioned for particulate reinforced metal matrix composites where void initiation occurs at sharp corner particulate stress concentrations.

The factors controlling the fracture toughness behavior of discontinuous metal matrix composites appear to be more complex. Not only is the toughness a function of reinforcement content, it is affected by both thermal and mechanical processing(8,10). These observations suggest that the fracture toughness of discontinuously reinforced metal matrix composites may be best understood by separating possible toughening mechanisms into their intrinsic and extrinsic components (35). Intrinsic components which control the toughness of these metal matrix composites include matrix heat treatment and the volume fraction, size and spacing of small intermetallic inclusions (typically less than 0.1 µm in size). The former will influence the ability of the matrix to relieve the stress concentrations associated with the propagation of a sharp crack by localized plastic deformation, while the latter will influence the composites' propensity towards void sheet formation.

Extrinsic components which control the fracture toughness behavior of discontinuously reinforced metal matrix composites include fiber size and alignment. For example, if a crack were propagating along the X-axis and encountered a particulate or short fiber reinforcement whose principal axis lay along the Y-axis, local crack deflection or crack bridging may occur,

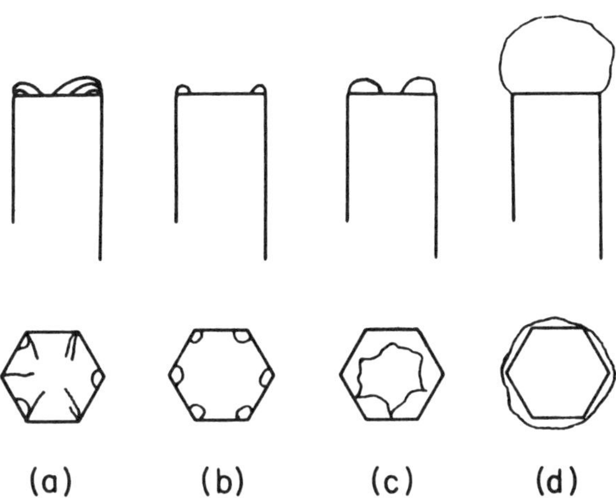

Figure 8 Schematic void inititation model for SiC whisker reinforced metal matrix composites(29).

the net effect being a reduction in the crack driving force. Theoretical calculations (36) also suggest that as the reinforcement spacing decreases, i.e., as the volume fraction reinforcement increases, the fracture toughness of the metal matrix composite should decrease. For two cylindrical inclusions, when widely separated, the location of maximum elastic stress concentration will be at the inclusion-matrix interface. However, as the spacing between the cylindrical inclusions approaches the inclusion size, the location of the maximum stress, and presummably the maximum strain concentration, shifts to the midpoint between the two particles. This analysis, if applicable to discontinuously reinforced metal matrix composites predicts that, as the volume fraction of constant sized reinforcement increases, the fracture path should change from one, at low volume fractions, which favors the reinforcement-matrix interface, to one, at higher volume fractions, which favors matrix fracture. Unfortunately no detailed fractograhic evidence has yet been presented which might completely confirm or refute this hypothesis. However, some support for this hypothesis is nevertheless given by You et al (32). Rather than seeking out reinforcement particulates, these authors' fractographic evidence suggests that crack propagation in 2124 aluminum reinforced with 20 volume percent α- SiC particulates, occurs randomly through the composite.

Acknowledgement

This research was initially supported in part under AFWAL/MLLS Contract F33615-82-C-5011, with Dr. T. Ronald acting as contract monitor. Subsequent studies of the fracture toughness behavior of discontinuously reinforced metal matrix are being supported by NASA-Langley under grant NAG-1-724, with Mr. W. Brewer acting as grant monitor.

References

1. "Application of Reinforced Metals to Cargo-Bomber Aircraft" (AFWAL-TR-3061, U. S. Air Force Wright Aeronautical Laboratory, June 1981).

2. A. P. Divecha and S. G. Fishman, "Mechanical Properties of Silicon Carbide Reinforced Aluminum," Proc. 3rd Int. Conf. on Comp. Mat'ls, Vol. 3, 1979, pp. 351-361.

3. D. Webster, "Effect of Lithium on the Mechanical Properties and Microstructure of SiC Whisker Reinforced Aluminum Alloys," Metall. Trans., 13A (1982) 1511-1519.

4. P. Parker and H. J. Rack, "Age Hardening Behavior of SiC Whisker Reinforced Al-Mg-Cu-Li Alloys," work in progress.

5. A. P. Divecha, S. G. Fishman and S. K. Karmaker, "Silicon Carbide Reinforced Aluminum-A Formable Composite," Jn. of Metals, 33 (1981) 12-17.

6. H. J. Rack, T. R. Baruch and J. L. Cook, "Mechanical Behavior of Silicon Carbide Whisker Reinforced Aluminum Alloys," Prog. in Sci. and Eng. of Composites, ed. T. Hayashi, K. Kawata and S. Umekawa (Tokyo: Japan Society for Composite Mat'ls, 1982), 1465.

7. H. J. Rack and J. W. Mullins, "Tensile and Notch Tensile Behavior of SiC_W Reinforced 2124 Aluminum," High-Performance Powder Aluminum Alloys-II, ed. M. Koczak and G. Hildeman (Warrendale, PA: The Metallurgical Society, 1986), 155.

8. T. E. Scott, J. W. Mullins and H. J. Rack, "Effects of Composition and Process Variables on the Properties of Discontinuous Silicon Carbide Reinforced Aluminum Metal Matrix Composite Materials (Report AFWAL-TR, U. S. Air Force Wright Aeronautical Laboratory, January 1987).

9. C. R. Crowe, R. A. Gray and D. F. Hasson, "Microstructural Controlled Fracture Toughness of SiC/Al Metal Matrix Comoposites," Proc. 5th Int. Con. on Comp. Mat'ls, ed. W. C. Harrigan, Jr., J. Strife and A. K. Dhingra (Warrendale, PA: The Metallurgical Society, 1985), 843.

10. W. H. Hunt, Jr. C. R. Cook, K. P. Armanie and T. B. Garganus, "Structure Property Relationships in P/M Aluminum Matrix-SiC Particulate Reinforced Composites," Powder Metallurgy Composites, ed. (Warrendale, PA: The Metallurgical Society, 1987), in press.

11. C. J. Skowronek, A. Pattnaik and R. K. Everett, "Dispersion and Blending of SiC Whiskers in RSP Aluminum Powders (NRL Memorandum Report 5750, Naval Research Laboratory, 1986).

12. S. R. Nutt and R. W. Carpenter, "Non-equilibrium Phase Distribution in an Al-SiC Composite," Mat'l Sci. and Eng., 75 (1985) 169-177.

13. H.J. Rack, "Age Hardening Behavior of SiC Whisker Reinforced 6061 Aluminum," Proc. 6th Int. Conf. on Comp. Mat'ls, in press.

14. G. Mott and P. K. Liaw, "Measurement of Mechanical and Ultrasonic Properties of SiC Metal-Matrix Composite," submitted to Metall. Trans., September 1986.

15. M. McKimpson and T. E. Scott, private communication with author, Michigan Technological University, September 1986.

16. D. Chellman, private communication with the author, Lockheed-California Company, October 1985

17. H.J. Rack, P. Hood, P. Nishanen and J. L. Cook, "Status of Large Sheet / Large Extrusion Development at Arco Metals-Silage Operation," Proc. 5th Discont. Metal Matrix Working Group (Santa Barbara, CA: Metal Matrix Comp. Infor. Center, 1983).

18. H. J. Rack and P. N. Nishanen, "Extrusion of Discontinuous Metal Matrix Composites," Light Metal Age, April 1984.

19. W. R. Mohn and G. A. Gegel, "Dimensionally Stable Metal Matrix Composites for Guidance Systems and Optics Applications," Advanced Composites: The Latest Developments (Metals Park, Ohio: ASM International, 1986), 69.

20. C. Standard, private communication with author, LTV-Vought, January 1987.

21. M. W. Mahoney and A. K. Ghosh, "Superplasticity in a High Strength Aluminum Alloy with and without SiC Reinforcement," Metall. Trans., 18A (1987) 653-661.

22. J. Ahern, C. Cooke and S. G. Fishman, "Fusion Welding of SiC-Reinforced Al Composites," Metal Construction, 14 (1982) 192.

23. H. L. Gegel, private commumication with the author, U. S. Air Force Wright Aeronautical Laboratory, 1983.

24. Y. V. R. K. Prasad, H. L. Gegel, S. M. Doraivelu, J. C. Malas, J. T. Morgan, K. A. Lark and D. R. Barker, "Modeling of Dynamic Material Behavior in Hot Deformation: Forging of Ti-6242," Metall. Trans., 15A (1984) 1883.

25. H. L. Gegel, J. C. Malas and Y. Prasad, "Thermomechanical Processing of Metal-Matrix Composite Materials," Proc. Tri-service Workshop on Dynamic Mechanical Properties, Characterization and Processing of Lightweight Metal Matrix Composites, ed. J. Bailey, S. Fishman and P. Parrish (Durham, N. C.: U. S. Army Research Office, 1984).

26. V. Nardone, "Fracture Toughness Behavior of Discontinuous Metal Matrix Composites," Proc. 9th Disc. Metal Matrix Comp. Working Group (Park City, Utah: Office of Naval Research, 1987).

27. S. V. Nair, J. K. Tien and R. C. Bates, "SiC-Reinforced Aluminum Metal Matrix Composites," Int. Metals Review, 30 (1985) 275-290.

28. D. L. McDanels, "Analysis of Stress-Strain, Fracture, and Ductility Behavior of Aluminum Matrix Composites Containing Discontinuous Silicon Carbide Reinforcement," Metall. Trans., 16A (1985) 1105-1115.

29. S. R. Nutt, "Interfaces and Failure Mechanisms in Al-SiC Composites," Interfaces in Metal Matrix Composites, ed. A. K. Dhingra and S. G. Fishman (Warrendale, PA: The Metallurgical Society, 1986), 157-167.

30. D. R. Williams, "Fatigue Crack Initiation"(Ph.D. Thesis, Northwestern University, 1985).

31. W. A. Logsdon and P. K. Liaw, "Tensile, Fracture Toughness and Fatigue Crack Growth Rate Properties of Silicon Carbide Whisker and Particulate Reinforced Aluminum Metal Matrix Composites," Eng. Fracture Mechanics, 24 (1986) 737-751.

32. C. P. You, A. W. Thompson and I. M. Bernstein, "Proposed Failure Mechanism in a Discontinuously Reinforced Aluminum Alloy," submitted to Scripta Metall., 21 (1987) 181-187.

33. P. K. Liaw, J. G. Greggi and W. A. Logsdon, "Microstructural Characterization of a Silicon Carbide Whisker Reinforced 2124 Aluminum Metal Matrix Composite," accepted for pub. Jn. of Mat'ls Sci., January 1987.

34. J. Coyne, private communication with author, Lockheed-Georgia Company, October 1986.

35. R. O. Ritchie and W. Yu, "Short Crack Effects in Fatigue: A Consequence of Crack Tip Shielding," Small Fatigue Cracks, ed. R. O. Ritchie and J. Lankford (Warrendale, PA: The Metallurgical Society, 1986), 167.

36. J. G. Goree, "In-Plane Loading in an Elastic Matrix Containing Two Cylindrical Inclusions," Jn. of Comp. Mat'ls, 1 (1967) 404-412.

Subject Index

Author Index